# How *Smart* Is Your **Cat?**

# How *Smart* Is Your Cat?

Discover if your pet can solve
these fun feline tests

## David Alderton

Quercus

# Contents

# Introduction

Cats have a reputation for not only being resourceful but also adaptable, so how quickly will your pet settle into a new home or learn new skills? Here's a unique guide to help you monitor your cat's progress and also to discover more about the lifestyle of your pet. This will benefit you both. By understanding more about your cat's behavior, you will be able to recognize when something is possibly wrong. The book also shows you how to build a strong bond of mutual affection and trust in the most effective way, and lets you measure your progress as you go.

Cats are intelligent and adaptable, living in our homes but keeping at least one paw in the wild. With a little investment in time and understanding, you'll soon start to discover just how smart your pet really is.

## AIMING FOR THE STARS

We are not talking pass or failure here. It's important to realize that all cats are individuals. Like us, they find it easier to master some tasks than others, and thus progress may be faster in certain areas than in others. So use this unique scoring system as a simple guide. Record your cat's stars on the charts at the end of each section and see the overall score on page 187.

★ = Getting to grips with the basics

★ ★ = Excitement as your cat displays new skills

★ ★ ★ = Ability and proficiency; your pet is a star pupil

Cats can be destructive around the home sometimes, but with some simple training, encouragement, and a ready supply of toys to play with, you should soon be having fun with a friendly, well-behaved, and clever feline companion.

Cats have a reputation for being totally independent, regarding our attempts to influence them with haughty disdain. But, in reality, it is possible to train a cat. By doing so, you should be rewarded with a pet who proves to be a much more sociable companion, being less inclined to damage the home or disappear if allowed to roam outdoors. You will need to be patient, however, to achieve the best results.

## UNLUCKY BLACK CATS

Researchers investigating how the coloration of cats may influence their behavior have found that black cats are generally the most docile, being happy to live together in urban areas. The irony, however, is that black cats remain less popular as pets because of their perceived links with witchcraft dating back over the course of centuries. Cat rescue organizations find them much harder to rehome.

Contrary to popular myth, cats are capable of forming very close bonds with people, soon coming to identify with individual family members.

however, did "wild" characteristics in the hybrid offspring become sufficiently diluted to ensure that these cats were relatively friendly and easy to live with.

**Left:** The wild instincts of cats still lurk just below the surface, explaining why they can easily revert back to a free-living lifestyle.

**Below:** Cats adapt well to many aspects of a domestic lifestyle. However, they do not develop any effective road sense.

## A MATTER OF BREEDING

An ever-growing number of distinctive cat breeds now exist. Aside from differences in their appearance, purebred cats also vary significantly in their behavior. The late 20th century saw great interest in developing breeds with striking patterning resembling the natural coats of certain wild cats. In some cases, crossbreeding between cats of this type and domestic cats was used. Not until the third generation,

# Getting to Know Your Cat

# Wild Instincts

Cats have a reputation for being independent spirits. This is largely a reflection of their evolution, as the wild ancestors of domestic cats were opportunistic hunters that had to rely on their skills and ingenuity in order to survive. The earliest remains of domestic cats discovered so far date back some 10,000 years. They were found on the island of Cyprus.

1 History suggests that cats chose to adopt us at around the time that agriculture was becoming established. Wild cats were attracted to grain stores where they hunted rodents, and humans soon began to encourage them to kill these pests.

2 It was in ancient Egypt that feline domestication began in earnest. A portrayal of a cat with a collar dating back some 4,400 years was discovered there. Cats ultimately achieved a divine status in Egyptian culture.

3 The wild cat (*Felis sylvestris*) is the original ancestor of domestic cats, although scientific evidence reveals that the Egyptians also kept and may have domesticated the jungle cat (*Felis chaus*) at an early stage.

Domestic cats retain many characteristics in common with wild cats, such as the ability to climb, and the two may interbreed in areas where their distributions overlap.

★ Your cat recognizes his name

★ ★ Your pet comes to you when he is called

★ ★ ★ He comes to you and "asks" to be petted

**Below:** Cat owners have observed that coat coloration can reflect temperament. Ginger cats are considered laid-back and friendly by nature.

**Above:** Domestic cats generally retain the keen hunting instincts of their wild ancestors. Never leave them alone with other small pets that they may view as their natural prey.

**Stripes**

Many of today's domestic cats still retain vestiges of striped tabby patterning, which typifies their wild relative. These markings serve to break up a cat's outline, helping it to merge more effectively into the background when hunting.

# Cat Characteristics

During the late 1800s cats began to be bred selectively for show purposes, and this led to the development of many recognizable breeds. As well as possessing a distinctive appearance, each breed also has a fairly definable temperament, which tends to reflect its ancestry.

The Ragdoll is justly considered one of the most gentle breeds. Its hunting instincts are significantly reduced.

1 Ordinary nonpedigree cats have a reputation for being easygoing, relaxed, and adaptable. Smart by nature and playful, they can adapt just as well to living with an elderly person as to being in a family with children.

2 Breeds like the American and British Shorthairs that were originally developed from so-called "street cats" of this type share similar traits, although they now exist in a wider range of colors.

**Age concerns**

It's not just the breed that is significant when considering a cat's character. All cats need to be properly socialized as kittens if they are to be friendly with, rather than nervous of, people in later life.

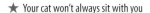

★ Your cat won't always sit with you

★ ★ He will sit with you when placed on your lap

★ ★ ★ Your pet willingly comes to sit with you

**3** All cats tend to be very alert by nature, as befits a hunter, but kittens in particular are very curious about the world around them and are constantly alert to new sights, sounds, and smells.

**Below:** A young Sphynx—this essentially hairless breed is only suited to living indoors as they are vulnerable not only to the cold, but also to sunburn.

**Right:** Bengal cats are hybrids, resulting from crossings of domestic cats with the wild leopard cat (*Prionailurus bengalensis*). As such, they generally have strong hunting instincts.

# Physical Attributes

When assessing the intelligence and character of a cat, you also need to take account of his physique, as this may exert an influence on his behavior. The way that the tail and legs in particular are configured is significant, as they affect a cat's ability to jump.

1 The Munchkin was originally highly controversial, because of its very unusual short legs. It was initially thought that this would be associated with a weak back, but happily this fear was misplaced.

2 Munchkins are just as healthy as their long-legged relatives. Rather than jumping, however, they are able to run straight under a piece of furniture when chasing a ball.

3 Munchkins and similar breeds are able to use their short legs just as effectively as a cat with legs of standard length. They do not need to be taught any specific behaviors.

The popular Munchkin is bred in short- and long-coated forms. A host of other emerging breeds now offer cats with similarly short legs.

### Cat creation
The most dramatic alterations in the physical appearance of cats have cropped up spontaneously in ordinary domestic cats. Breeders have then sought to "fix" these characteristics, creating breeds with features such as short legs, folded ears, or missing tails.

★ Your cat is active and runs around a lot

★ ★ Your pet is always keen to play with you

★ ★ ★ He seeks you out to play a game with him

**Left:** The Persian generally has a very placid nature, and is well-suited to living indoors. He is far less inclined to wander than some breeds such as the Norwegian Forest Cat.

**Right:** Lithe and agile, the Siamese and its close relatives making up the Oriental group are talented climbers, able to scale trees outdoors or drapes in the home with equal ease. They are also very vocal.

# Scaredy Cat!

**N**ever forget that cats are much more nervous by nature than dogs. Their instinct means they are always alert to danger because in the wild their small size leaves them vulnerable to larger predators. Just as cats rely on their acute senses to detect possible prey, they also rely on them as an early warning system to warn of any likely threat.

1 The acute sensitivity of a cat's hearing means that loud noises can be very upsetting, especially if accompanied by repeated bursts of bright light.

2 Cats may hide away under these circumstances, remaining out of sight somewhere they feel secure, such as under a shed or in a garage.

3 Some cats are more nervous than others and, if frightened, they may even go into hiding for several days before eventually returning home.

Always get your cat indoors when there are likely to be firework displays, or when thunder and lightning have been forecast.

★ Your cat gets very distressed by loud noises

★ ★ He generally ignores them if the drapes are drawn

★ ★ ★ He settles down and sleeps despite the noise

### Know the signs
A scared cat draws his ears back, hisses, and raises his fur as shown in the picture below, so as to appear more intimidating. But take care when you see these signs, as he may scratch or bite you while in this agitated mood.

Kittens scare particularly easily because of their lack of experience of the world.

# Small Spaces

**M**ost cats like to hide away in small spaces around the home from time to time, often in darkened surroundings. Do not be surprised if your cat curls up in unlikely places in the house, even in a drawer or on the top shelf of a warm airing cupboard.

1 Cats were often a welcome part of a ship's company on long sea voyages in past centuries. They were a convenient means of controlling the rat and mouse population onboard.

2 The ancestors of today's Maine Coon breed gained a reputation for sleeping in the most unlikely spaces in these nautical surroundings despite their relatively large size.

3 Cats are able to curl into a tight ball when they sleep, particularly if the temperature is cold, but will then stretch out if they become hot.

Cats like this Maine Coon are always on the look-out for possible new sleeping places around the home.

**True or False?**
Teacup kittens are so-called because they will seek out teacups in which to sleep. (Answer on score chart, p76)

☐ True
☐ False

**Left:** Cats may sometimes doze off in the sun lying on a branch or a wooden beam as here, but remarkably they rarely fall off.

**Below:** Sometimes a cat's quest for unusual sleeping places leads him into difficulties and he becomes trapped. Keep this in mind if your pet suddenly disappears from the home.

**Why do it?**
Domestic cats instinctively prefer enclosed spaces for sleeping purposes, having inherited this trait from their wild relatives. They may even seek out a den under the roots of a tree where they feel relatively safe.

# Motivate Your Cat

A key difference between cats and dogs is that cats are less inclined to do things to please their owners. Yet with the right encouragement, there is no reason why you cannot persuade a young kitten to adapt well to your routine and enjoy your company.

Never leave a young kitten alone with a dog, as your canine companion may not be entirely happy about the arrival of the newcomer.

## GETTING STARTED

1 With a good selection of toys to choose from, a kitten will soon learn the rules of the game, whether pouncing on a pull toy or chasing after a ball.

2 Play is training for hunting. Kittens are not born with innate hunting skills, but you will soon notice that your pet's coordination and skill will improve as he plays imaginary hunt games.

3 Kittens have high energy levels, but generally they will only play in short, intensive bursts before falling asleep or looking for food. This is quite normal behavior.

★ Your cat "asks" you to be fed

★ ★ Your pet recognizes his set mealtimes

★ ★ ★ He comes indoors at the right time for a meal

### Nighttime wandering

Wild cats become active as dusk falls, preferring to sleep for long periods during the day. Domestic cats may display similar behavior and so you need to train your cat to come in at night, so he will be safe and sound indoors.

**Above:** Set aside regular times each day to make a fuss of your cat, so he will get used to being picked up and stroked.

**Right:** Establishing a regular feeding routine is vital, so that your cat learns when mealtimes are, rather than pestering you for food at random times of the day.

# Learning to Hunt

**I**n the wild, hunting prey is the most important lesson that a young kitten needs to master, and he learns the requisite skills by watching his mother. Once they are old enough, kittens will accompany their mother on hunting expeditions, watching her in action.

1 A cat's vision is vital to his hunting abilities. Cats possess binocular vision, meaning that the visual input from each eye overlaps to an extent, creating a very precise central area of the image that the brain registers.

2 This image gives a clear indication of where the prey animal is positioned, enabling the cat to pounce with great accuracy. To maximize chances of success, the cat must get as close as possible without being detected.

3 A very stealthy approach, using whatever cover is available, is therefore essential, and explains why cats move very slowly and cautiously, watching their prey closely, with the aim of getting close to their target before launching a strike.

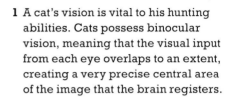

Kittens like the sound of the bell that is concealed in some toys. Your pet may soon learn to play by himself, patting the ball away and then leaping on it.

Cats like seizing knitted toys in their claws and biting them, but be sure they cannot accidently swallow threads or become tangled up in them.

★ Your cat starts to adopt a hunting posture

★ ★ He times his leap but misses the moving target

★ ★ ★ Your cat pounces and catches the target

### Killer instinct

Cats sometimes appear to toy with creatures they catch, probably because although their hunting instinct is strong, they have not learned how to dispatch their prey. Kittens reared on farms are generally the most effective hunters.

# Weather Forecasting

Can cats predict the weather? There is a long-standing belief that they may possess some weather-forecasting skills, and even evidence to back up such claims. But don't rely on him—it can be a bit hit-and-miss!

1 Some breeds, such as the Norwegian Forest Cat, are much hardier than others and so will be less affected by bad weather. They continue to roam and hunt outdoors in the rain.

2 During cold weather, expect your pet to spend longer indoors in the warm. During very cold, snowy spells, your cat may be reluctant to set foot outdoors at all.

3 Your cat is also likely to seek out shelter indoors if there is a storm in the vicinity, although many cats do not mind getting wet.

Belief in cats as weather forecasters often stemmed from the experience of sailors at sea. If a ship's cat became boisterous, repeatedly wagging his tail, this was regarded as a sure sign of gales on the way.

- ★ Your cat's behavior sometimes changes unexpectedly
- ★ ★ Changes occur in parallel with weather variations
- ★ ★ ★ You can predict the weather from his behavior

### Storm alert

It is possible that a cat's acute hearing allows him to pick up the sounds of an approaching storm long before such indications are audible to our ears. These distant sounds can cause your pet to start behaving strangely, appearing both excited and nervous at the same time.

**Above:** The behavior of young kittens is usually unaffected by the weather, since, even if they are allowed out, they will not be likely to venture far from the safety of home.

**Left:** If a cat cannot get back indoors during a storm, he is likely to hide away in an outbuilding or similar place of shelter, only emerging once the storm has passed.

# Nature or Nurture?

There is no doubt that the individual personalities of cats do differ, and quite markedly so. In terms of the overall domestic cat population, the way that individuals have been brought up is more significant than their origins, as comparison of the behaviors of domestic cats and strays reveals.

1 Cats revert back to living wild very quickly, and can form colonies in some areas, typically in old industrial sites where there is lots of cover and prey such as rats is plentiful.

2 Even very young kittens taken from such surroundings do not settle well as pets. Therefore animal welfare organizations trap and neuter these so-called feral cats, releasing them afterward rather than trying to domesticate them.

3 This neutering program means that the colony will no longer continue to grow in numbers, but the group can remain living together. Neutering also lessens the risk of outbreaks of fighting, which can lead to injury.

Some breeds such as the Abyssinian are regarded as being more adaptable and child- and dog-friendly than others.

★ You have to seek out your cat for company

★ ★ He comes to you when it is time for his dinner

★ ★ ★ He greets you when you return home

### Rescue me!

Although most people enjoy the fun of homing a new kitten, there are always lots of older cats living in cat rescues that are in need of permanent homes. These individuals can make great companions, so do consider the option of adopting a more mature pet.

**Above:** Cats soon come to recognize their owners, and develop a close bond with them. Establishing a daily routine is important to help a cat integrate into the household, and settle down well.

**Right:** A purebred show cat that has spent much of his life previously in a cattery will need time to adjust to living in domestic surroundings.

# Basic Instincts

Cats remain remarkably self-reliant, with their innate hunting instincts still being apparent in their day-to-day patterns of behavior. Whether stalking a plastic bag blowing in the wind, running along a fence, or pouncing on a ball, all such behaviors are a reflection of a cat's basic hunting skills.

## PATIENCE PAYS OFF!

1 Perhaps the most crucial hunting skill that a cat has to master, through repeated play, is finding the best way to approach his target. He must creep up as close as possible to his potential prey without being spotted.

2 You can usually evoke this behavior quite easily in a young cat by using a stout piece of rope as a lure. Drag this along the ground, preferably through grass, and watch how your cat is drawn to follow it as it moves.

### Practice makes perfect

At first, young kittens do not have the necessary coordination to hunt very effectively but, before long, their senses sharpen and they will be able to leap up in the air to grab a passing leaf blowing in the wind.

A young kitten displaying stalking behavior. Part of this is instinctive, but watching how his mother hunts will reinforce this response. Young cats also edge forward warily in this way when they are slightly nervous.

**3** Cats use their keen sense of sight to follow a target, and will respond to any changes in its pattern of movement. If you stop pulling the rope along after a time and pause, watch how your cat freezes in response.

**Right:** Staying low and moving slowly and quietly are strategies employed by cats to avoid detection when stalking. Finally, when close enough, the cat pounces.

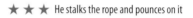

★ Your cat watches if you pull a length of rope along

★ ★ Your pet follows the rope as you pull it for him

★ ★ ★ He stalks the rope and pounces on it

**Left:** The shape of a cat's eyes can vary from round, as in this Singapura, to more elliptical forms. They are vital for coordination. A cat can see better in the dark than us, and is able to judge distance with great precision.

# Seasonal Games

Cats will invent all sorts of games, chasing after and pouncing on moving objects, in the absence of real prey. Although your pet will enjoy playing with you, do not be surprised to see him following his own instincts and amusing himself outdoors.

1 Anything that blows around your yard is likely to attract your cat's attention. Leaves and lightweight trash such as pieces of paper or food packaging will be incorporated readily into play routines.

2 In the summer, your cat will spend longer outdoors, sleeping in cool areas under bushes when the sun is hot. Just watch out if he is inclined to chase after and catch flying insects such as bees—they can sting.

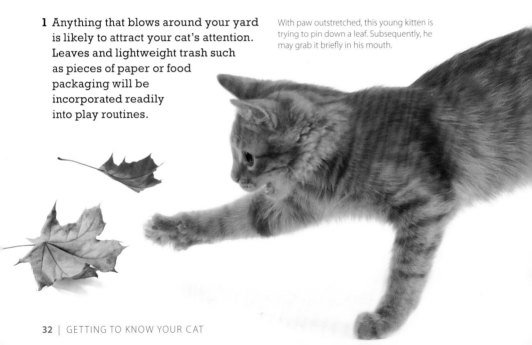

With paw outstretched, this young kitten is trying to pin down a leaf. Subsequently, he may grab it briefly in his mouth.

**Snow fun**

Kittens encountering snow for the first time can react quite excitedly, running around and trying to catch the snowflakes before they hit the ground.

★ Your cat likes playing with you outdoors

★ ★ Your pet chases dry leaves blowing in the wind

★ ★ ★ He catches them in his paws or mouth

**Right:** It's not all about play! A comfortable dry bed of leaves is a great location for a snooze.

**Right:** Ripe fruit in the fall equals instant game! Cats love to pat and play with windfalls. Apples are better than pears because they are rounder and roll more easily.

# Up Close and Personal

Cats express affection for their owners in some very obvious ways, and in some which are perhaps a bit less obvious . . . The signs that your cat likes you reflect the way in which cats communicate with each other. It does, however, take time for your pet to build up a deep bond with you.

1 The most obvious way in which your cat recognizes you will be by a combination of your appearance and the sound of your voice. Cats, like dogs, can be trained to come to you when called by name.

2 Cats also use their own scent (left by rubbing against you) to identify you as their owner. But this is essentially used as a marker to highlight their association with you to other cats.

> **Cupboard love**
>
> Once your cat has bonded with you, he will start to rub his head against your legs, particularly when seeking food. By behaving in this way, he is actually depositing his scent on this part of your body. Although totally indiscernible to us, other cats will recognize this marker.

This cat is used to being picked up and is not struggling. Generally, a bit more support around the hindquarters will help a cat to feel relaxed when being held.

**3** Vocalizations are also an important part of the way in which a cat communicates with his owner. A cat may meow persistently when he wants to be fed, although this depends to a great extent on the individual.

★ Your cat lets you stroke him

★ ★ Your pet comes when he is called

★ ★ ★ He looks out for you and comes when you return

A friendly cat will come and find you both for company and also when he wants to be fed. Cats that are well socialized and not nervous of people may even respond to strangers out on the street.

# Aging Gracefully

Cats are adept at concealing their advancing years. The signs of aging are quite subtle, and most cats remain active right up until the very end of their lives. Thanks to advances in health care and a better understanding of their nutritional needs, domestic cats are living longer than ever before.

1 Most cats today have an average life expectancy which extends into their late teens or early twenties. Unlike dogs, any black areas of fur around their jaws and eyes don't turn gray as they become older.

2 An obvious sign of your cat getting older will be a reluctance to sit on your lap. The position becomes uncomfortable for them and he will probably prefer to rest just next to you.

Older cats may suffer from stiffness and not be able to groom themselves as effectively as they once could. They are also more likely to suffer dental problems and constipation as they grow older.

## Changing the diet

There are now special foods that have been formulated to meet the changing nutritional needs of older cats. They are particularly vulnerable to a decline in kidney function, which can lead to bad breath (as tartar on the teeth also does) and weight loss.

**Below:** It can be difficult to age an adult cat accurately, particularly from middle age onward. The condition of the cat's teeth is a helpful indicator for someone with the right experience.

**Right:** Older cats are likely to be less active than when they were younger, but they still like to seek out warm spots where they can snooze in the sunshine.

**True or False?**
The oldest cat on record lived for 38 years.

☐ True
☐ False

# Popular Breeds

An ever-growing range of both breeds and color varieties is now available, and the choice of a purebred feline companion has never been greater. Certain traits are associated with different breeds, and their characteristics tend to be more tightly defined than is the case with a nonpedigree cat. People are also drawn to particular breeds because of their distinctive appearance. There has been a noticeable trend over recent years to create new breeds that resemble miniature wild cats, with striking coat markings. In some cases, as with today's Bengal breed, initial crosses with wild cats were used to achieve this particular goal.

Today's Siamese cats are characterized by their large ears, triangular-shaped head, and blue eyes, combined with a pale body that is marked at the extremities.

**Left:** With long-haired breeds, the coats of kittens are always shorter than those of adult cats. You must be prepared to spend a lot of time every day grooming your pet once it grows up.

**Right:** Short-coated breeds such as the Abyssinian just need normal grooming. Here selective breeding has removed the tabby barring, and the result is a ticked tabby.

### Did you know?

On the whole, pedigree breeds are likely to live just as long as ordinary crossbreed cats. Fashion does play a part in the popularity of breeds, however. If you are interested in a particular breed, always check out its temperament and care needs in detail, before finalizing your choice of cat.

# Teaching
# the Basics

# The Name Game

Pet cats around the world are known by countless thousands of different names. People always seem to be a bit more imaginative when choosing a suitable name for a cat, compared with one for a dog. If you are stuck, there are even books to help you. But don't get too anxious, cats themselves are not fussy!

1 Consistency is the key, whatever name you decide on. It's important to get your cat used to the sound of her name as soon as possible.

2 Use your cat's name whenever you need to call her, and particularly when it's time for her food. She will soon learn who you mean and will come running to you!

3 It is best to choose a distinctive name that cannot be confused with those of any other pets that you own. Cats have very sensitive hearing, which helps when you are calling them.

The zoological name of the wild cat ancestor is used for cat breeds created by hybridization. The Bengal is the best-known example, being named after the leopard cat (*Prionailurus bengalensis*).

★ Your cat seems to ignore her name when you call her

★ ★ She comes to you indoors when you call her name

★ ★ ★ She comes in from outside when called

### What's in a name?
Try to find out the name of a cat that you are rehoming as soon as you get her. Using her familiar name should help her to bond with you more quickly.

**Above:** Always use your cat's name whenever you are talking to her, so she comes to recognize the sound and responds to it when you call her.

Cats can be trained to come when called by their name, especially if rewarded with a treat.

# Litter Training

Training your pet to use a litter tray is a very important aspect of cat care, especially when you have a house cat who lives permanently indoors. Fortunately, cats are instinctively clean animals and it is usually much simpler to housetrain a cat than a dog. In fact, most kittens will use a litter tray readily from the time of weaning onward.

1 Be aware—kittens are most likely to want to relieve themselves when they wake up and after eating, so move them to the litter tray at these times.

2 Cats have sensitive noses. One of the most common reasons why they will ignore a litter tray is because it is dirty, so keep the litter clean.

3 Place the tray in a quiet yet easily accessible part of your home. Stand it on some sheets of old newspaper to catch any litter scraped out of the tray.

Use a special scoop to minimize the amount of cat litter that has to be thrown away when the tray is soiled, and choose a cat litter that clumps when wet.

**Above:** Cats like privacy and some prefer to use a hooded litter tray. Some designs also trap unpleasant odors if you are not around to clean the tray within a short time of it being used.

★ Your cat investigates and sniffs the litter tray

★ ★ Your pet is happy to stand or sit on the litter

★ ★ ★ She uses the litter tray for its intended purpose

Odor-absorbing types of cat litter are also available, along with liners for the litter tray that make it easier to clean the tray thoroughly. They can be simply lifted out and disposed of in one piece.

# Carpet Digging

It can be very annoying if your cat comes indoors, and then straight away starts to scratch at the carpet. Such behavior is often described as "stropping"! Cats tend to use the same area when they are clawing like this, so it won't be long before she has inflicted noticeable damage on this spot.

1 The health of their claws is a very important consideration for cats, since these serve as an essential part of their armory. They help your pet to hunt and climb effectively.

2 There are various ways of deterring unwanted scratching at the carpet. You can use a deterrent spray, or simply keep a water pistol handy and squirt your cat if she misbehaves in this way.

### Scratching about

A scratching post can be recommended for this problem, especially one treated with catnip, which should immediately serve to attract your pet's attention. Scratching here will encourage her not to do it randomly elsewhere indoors.

**3** If the carpet is being scratched next to a door, it may be that your cat is feeling trapped in that room and wants to leave. Her scratching indicates that she is trying to dig herself out.

**Above:** If your cat starts to get caught up by her claws on a scratching post, it suggests that they may be overgrown and require trimming back slightly.

**Right:** Your cat usually keeps her claws retracted when walking about the home. They will only start to snag and catch on the carpet if they are overgrown.

★ Your cat sometimes scratches at the carpet

★ ★ Your pet will sometimes use a scratching post

★ ★ ★ She always uses a scratching post, never the carpet

# Establishing a Routine

The lives of most pets, much like our own, rely on the establishment of a set routine. Cats are no different. A kitten wakes up, she needs to relieve herself, she is hungry and wants to have food and a drink, followed perhaps by a game and then a period of sleep.

1 Encouraging your cat to develop a routine is sensible because you can then effectively merge your lifestyles together in a harmonious way. Train your cat to come in at night, rather than expecting to go out then.

2 The routine should not just cover occasions like mealtimes, but should also apply to other aspects of care, like handling. Regularly picking up and stroking a young kitten will help her to become more affectionate.

3 Cats generally adapt very well to gradual alterations in domestic routine. However, the presence of unfamiliar people in your home, such as builders, can be upsetting and cause your pet to hide away.

Be prepared to carry out some adjustments to the routine as your pet grows older. For example, adult cats require fewer, but larger, meals in the course of a day than kittens do.

**True or False?**
Cats invariably follow a routine before settling down to sleep, by circling around their chosen spot.

☐ True
☐ False

**Above:** Playing with your pet each day will deepen the bond between you, as a routine develops. Before long, your cat will seek you out to play games with her.

**Below:** Being regularly groomed from an early age means that your cat will look forward to this attention as she grows older.

**Hidden routine**
Your cat's routine will also be influenced by familiar scents in her environment, which you will not be aware of. Changes in these markers can have dramatic effects on a cat's behavior, even triggering a breakdown in toilet training.

# Day to Day

**A**s your cat settles in with you, she will start to recognize key domestic events in your daily routine. Her reaction to your activities may depend on whether they are likely to disturb her or not. If you are simply going out for a while, your pet will probably be totally unfazed and just carry on sleeping.

1 Anything that involves loud noise is likely to draw a swift response from your cat, as they prefer quiet surroundings. She will soon recognize when you are about to vacuum around the room, and beat a retreat.

2 Simply seeing the vacuum cleaner may cause your pet to get up and head off to another part of the house—or even ask to go outdoors—before you start using the machine.

Cats dislike the noise of washing machines but they may see a basket of freshly dried laundry as a warm and inviting opportunity to jump in and doze.

**3** Cats will often appear at your feet when you are cooking, in the hope of being given a tidbit. Keep an eye on them—you don't want them to jump up and burn their feet on the stove.

### Patterns of behavior
Your behavior will influence that of your cat, and over time a daily routine will develop. These routines may be influenced by the cat's innate curiosity.

★ Your cat watches what you are doing around the home

★ ★ She actively reacts to your daily routine

★ ★ ★ Your pet develops a routine alongside yours

**Above:** Anything that is food-related may arouse your pet's curiosity, such as bags of food brought back from a shopping trip.

**Below:** Cats often like the warmth of a bathroom and they may decide to join you while you are having a relaxing soak in the tub.

# A Day in the Life

Once your cat has settled in as part of your family, it's quite likely that she will start to follow you around when you are carrying out everyday tasks, seeking out your company. This is not something that you can plan, of course, but it does reveal the development of a deepening bond with your pet.

1 Cats are great observers—this reflects their hunting methods and their general lifestyle. They will soon become attuned to your routine.

2 They are particularly likely to identify the aspects of a daily routine which involve behaviors that benefit them, such as feeding times.

3 If your cat feels that something in your routine has changed, she may actively start trying to attract your attention— by rubbing against you, for instance.

**Below:** Your cat may join you when you begin doing exercises on the floor, but she is unlikely to start revealing a taste for yoga!

**Below right:** When you are sitting down quietly, your pet may join you. Place a blanket on the couch to gather up your pet's loose hairs.

A watchful cat will soon figure out when you leave for work or go shopping each day. There is also evidence to show that your pet will anticipate when you are likely to return.

★ Your cat sometimes sits alongside you

★ ★ Your pet regularly shares the couch with you

★ ★ ★ She gets up to greet you when you come home

# Cat Flap Challenge

A cat flap is a useful piece of kit, especially if you are out during the day, because it lets your cat wander in and out of the home at will. The potential drawback, however, is that if a problem develops either indoors or outside, you will not be there to intervene.

1 There are various types of cat flap. Ideally, choose a magnetic one that only allows your cat to enter. This prevents other neighborhood cats from coming in, stealing food, and potentially soiling your home.

2 With a bit of encouragement your cat will soon learn to master a cat flap. The first step is to persuade her to step through the opening on her own, without the flap in place. Then you can progress to having the flap down.

★ Your cat goes through the opening without the flap

★ ★ Your pet is able to push through the flap

★ ★ ★ She goes in and out of the flap at will

The cat flap was invented by the scientist Isaac Newton (1643–1727), who was a great cat-lover and wanted to improve the quality of his pet's life.

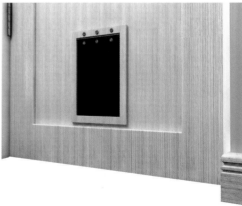

**Not upfront!**
Don't be tempted to situate the cat flap in your front door, so it opens onto the street. This will put your cat at greater risk of being hit by a car if she runs out on to a busy road.

Hold the cat flap up at first, so your cat will learn more quickly that she can step through the open hole. Always reward your pet when she responds like this.

# Clickety Click!

Clicker training has now become very popular for training a wide range of animals, including cats. You can obtain the clicker easily from pet stores or online. There may be slight differences in design, but they all operate in a similar manner.

## USING A CLICKER

1 It is the sound of the device—the click—that is significant when training your pet. Cats, who have sensitive hearing, will soon pick up on this noise.

2 During training, keep the clicker with you at all times. Then, when your cat responds in a way that you like, press it immediately and occasionally reinforce the positive behavior with a tasty treat.

Your cat will soon come to realize that the sound of the click is an indicator that she is doing something that may bring a more substantial reward, such as a treat.

**3** Cats, which are naturally curious, may be initially distracted by the sound of the clicker but it does mean that your pet is registering its distinctive sound.

**4** This method of training is sometimes described as "positive reinforcement," with the noise of the click helping to reinforce that particular action in the cat's mind.

★ Your cat responds to the sound of the clicker

★ ★ She starts to pick up the routine when you click

★ ★ ★ Your pet responds positively to the clicker

### Starting to click!
Over time, the clicker can help you to build up a more complex routine with your pet. This can be done in sections, by linking steps together. But it is not a magic solution to having a well-trained cat!

Be prepared to devote enough time to training your cat properly. Some pets learn faster than others. The clicker is a useful communication aid in the training process.

# Auto-Feed

Automatic feeders operate by battery rather than mains power, so even in the event of a power cut, your cat will not be deprived of food.

**C**ats can be very fussy about their food. If your pet does not like what you are providing or simply feels hungry, she may well head off in the direction of a neighbor's home for a meal. An auto-feeder will ensure that your cat gets her meal at the usual time, even if you can't be there to feed her.

1 Providing fresh food is vital. You can simply leave dry food out for your cat when you are out, but this is not something that all cats will eat readily.

2 Wet food needs to be kept chilled and covered to prevent it drying out in a warm atmosphere and attracting flies, which can be carriers of disease.

3 Cats soon master how an automatic feeder works, but be there to supervise initially, so you can help your pet understand the way in which the cover opens to reveal her food.

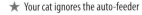 Your cat ignores the auto-feeder

★ ★ She will eat from the feeder when the cover is open

★ ★ ★ Your pet sits waiting for the feeder to open

### Waiting for dinner

Automatic feeders are ideal for regular use, as cats rapidly learn how they operate. Before long your pet will be sitting by it, waiting for the cover to spring open, as the final seconds tick away before her meal is served.

**Left:** Auto-feeders are less suitable if you have two cats, as the dominant individual may eat all the food, leaving her companion hungry.

**Below:** Stand the feeder on linoleum or tiles, rather than carpeting as here, in case some food is spilled. Always wash the bowl after use.

# Playing Ball

Cats enjoy playing ball, using their remarkable coordination for this purpose. Games of this type provide a great way to establish a bond with your pet, and encourage a cat to take exercise. Irrespective of their age, all cats generally enjoy this type of fun activity.

Always use soft balls to avoid any risk of your cat being injured. They are also easier for your pet to grasp and bat around.

1 It is simple to start a game. Just roll the ball across a level surface to get your cat's attention. At first she may just watch from a distance rather than actively chasing the ball.

**The right ball**

Choose a suitable ball for your pet, and make sure that it is lightweight and cannot cause her injury. Kittens will prefer to play with a smaller size ball. As they grow, you can introduce larger ones.

**2** Before long, your cat will happily chase after the ball and pounce on it. Some cats crouch down and like to ambush the ball as it rolls past them.

★ Your cat watches the ball rolling along

★ ★ She starts chasing the ball when you roll it

★ ★ ★ She starts playing with the ball by herself

**3** Once your cat is used to playing with a ball, she is likely to start a game all by herself. This may partly be in the hope that you can be persuaded to join in the game—so go on and have fun!

**Below:** This type of toy rolls around like a ball but has the added advantage that a cat can pick it up easily.

**Right:** The floor surface affects the game, as a ball will run more quickly over a wooden floor. But your cat will be more likely to slip and slide while chasing it across a slippery surface.

# Stalk and Pounce

Stalking and pouncing mimics a cat's typical hunting behavior. These are physical actions that your pet is likely to display during play. This type of response can be seen both in young kittens and older cats, although it is not something that you can teach. These behaviors are instinctive.

1 The best way to persuade your cat to stalk and pounce is to drag a target lure, such as a length of thick string or a l ribbon, across the ground to attract her attention.

2 Your cat will follow the string, moving slowly and deliberately after it. But if you stop pulling, your pet will probably stop in her tracks as well.

3 If your cat is sufficiently confident, she may leap on top of the string at this stage, after having watched it intently as if she were hunting prey.

★ Your cat watches a piece of string being pulled along

★ ★ Your pet starts to follow and stalks the string

★ ★ ★ She pounces on the string and catches it

Under normal circumstances, wild cats stalk prey using whatever natural cover is available, so they can spring an ambush. This means they can get close to their target before striking, increasing the likelihood of success.

## Stealthy hunters

Cats are natural predators that rely on stealth to get as close as possible to a target. This greatly increases their chances of catching prey unawares.

**Above:** Cats will display stalking behavior when hunting birds in a garden, flattening their bodies to the ground and creeping forward slowly.

**Left:** This kitten is striking out to grab the toy. Note her exposed claws which she will use to grab the feathers.

# Cat Agility

When jumping down to the ground, a cat touches down on her front feet first, using the so-called "stopper pads" on the back of her paws to act as a brake.

Cats are among the most lithe and agile of mammals and these attributes stand them in good stead as solitary hunters in the wild. They can climb well, run along branches, and jump back down again to the ground without difficulty. Even if they are unfortunate enough to slip from a height, they can swivel their body round in a fraction of a second to land on their feet, greatly reducing the risk of injury as a consequence. On the ground, they can crouch down low or leap up high into the air, emphasizing their natural suppleness and athleticism.

**Left:** If a cat falls from a height of more than about 20 feet, she is likely to suffer a fractured jaw as the result of her lower jaw hitting the ground when she crash-lands.

**Below:** Cats often jump on to an outbuilding or tree in order to get access to the top of a fence. Most fences are too narrow for a cat to reach directly from the ground so it helps to have a higher perch from which to step onto it.

**Did you know?**
Domestic cats are formidable jumpers and are able to leap a distance of around 8 feet without difficulty. This means that they can easily reach a tabletop or work surface from a sitting position on the ground.

# Essential Grooming

Cats of different breeds differ quite widely in the amount of grooming that they require. Long-haired cats need a lot of regular grooming to keep their coats free from tangles and to ensure that they look at their best.

1 Not all cats enjoy being groomed, especially if they have had a bad experience in the past. This is why it is a very good idea to train your cat that grooming is nothing to fear.

2 The coats of young long-haired cats are shorter than those of adults, which makes it easier to accustom them to being groomed from an early age.

Be careful not to pull at your cat's coat when combing her fur, as this will be painful for your pet.

**Fur balls**

Cats are at risk of getting fur balls as loose hairs get stuck to their tongues as they groom themselves and end up being swallowed. This can lead to a blockage in the stomach that causes the cat to go off her food.

★ Your cat struggles when being groomed

★ ★ She lets you groom her but tends to walk away

★ ★ ★ She is totally relaxed about being groomed

Once she is used to being groomed, your cat will enjoy the experience and will lie in a relaxed fashion purring loudly while you brush her.

**3** Choose a comb with rotating teeth as this helps to untangle knotted hair. If the fur is matted, however, it may be less painful to cut the mats out, and let the coat grow back over time.

Be sure to choose the right grooming tool for the job. These wire slicker brushes should always be used with care, especially if your cat is nervous or unused to being groomed.

# Fussy Feline

Cats have a well deserved reputation for being very fussy about food, but a cat's reluctance to eat may not always be due to the food itself. It can also be related to the manner in which they are being fed and even the choice of room where you put down the food bowl.

## FEEDING TIME

1 Most cats prefer wet food, whether in cans or sachets, rather than dry kibble. However, they are often not keen to eat food straight out of a fridge as it may be too cold for them to enjoy.

2 Always make sure that your cat's food bowl is clean. Any lingering odors of stale food are likely to put your cat off eating from it.

3 Choose a quiet spot in the home to feed your cat, rather than a place where people are bustling around, particularly if this is a noisy area too.

Offer different types of food to a kitten so that she gets used to being presented with various options. This helps to prevent a reluctance to eat when she is given something unfamiliar later in life.

**Be patient!**
Cats with nasal infections often lose their appetite because their sense of smell is compromised. After illness, a cat may demand a change of food too. After two or three months, you can trying offering the original food again, as your cat may then be happy to eat it.

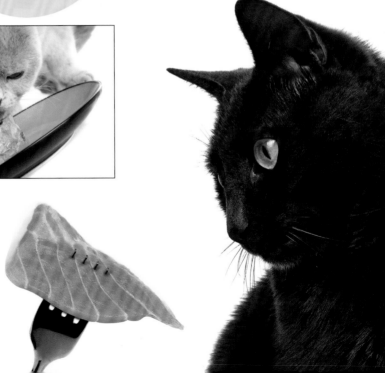

**Above:** Cats rely heavily on their sense of smell when it comes to determining whether they will eat an item of food or not.

**Right:** Cats naturally are hunters, not scavengers, so warm food appeals more to them because it is similar in temperature to freshly killed prey.

# Food Obsession

Although cats can often be fussy about their food, they can also become quite obsessed with various foods that we eat. If you feed your pet tidbits from your plate when you are having a meal, beware! You could easily find yourself being pestered again in future whenever you are eating.

1 Cats are creatures of habit but they also learn very rapidly if something is to their advantage. Their opportunistic nature makes them alert to any chance of getting a tasty treat.

2 If you get into the habit of feeding your cat at the table and then suddenly decide to ignore her in an attempt to break the habit, you may find that she jumps up on the table to see where her missing treats have gone.

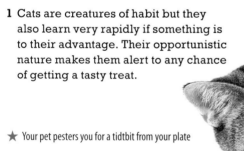

Cats will steal food if they have the opportunity. They can easily leap onto a table or kitchen worktop and simply help themselves.

★ Your pet pesters you for a tidtbit from your plate

★ ★ Your cat joins you when you are eating a meal

★ ★ ★ She doesn't pester you at mealtimes at all

**Above:** Cats are well equipped to steal food off a work surface. They can use their long legs and claws to drag food toward themselves before seizing it with their teeth.

**Left:** Cats definitely prefer meat-based foods, even if these are raw. They may sample other foods too, such as cereals and milk as shown here.

# Little Beggar

**C**ats can be very inventive when on the prowl for a tidbit or treat. Although dogs are better known for begging, this cute trick is also something that you can teach a cat without difficulty, quite often in a relatively short space of time.

1 A cat's ability to beg partly stems from her well-muscled hindquarters, which provide the power that allows the cat to support herself when sitting with her front paws raised.

2 The tail helps to stabilize and support the cat by providing additional balance for her body. Cats also sit in this way when trying to reach up for something from an elevated surface.

> **Beg to differ!**
> While kittens can be taught to beg readily, simply by holding up a reward, this response is reduced in older cats, often because their bodies are no longer as supple.

**3** Be careful when offering your cat a treat in this position, because she may swipe at you with a paw and unsheathed claws in an attempt to knock it out of your hand.

★ Your cat sits on the floor looking up at the treat

★ ★ She sits up on her hindquarters to reach it

★ ★ ★ Your pet begs readily when offered a treat

A cat will use her front feet and claws to reach a tidbit when sitting upright, particularly if it is just out of reach. The position in which you offer the tidbit is important. Don't hold it too high—it needs to be within easy reach.

# High Five!

This trick can be taught most easily to young kittens, because they tend to learn more readily than older cats. Take care! You need to be watchful that, in her exuberance, your cat does not inadvertently scratch you. Cats—and particularly kittens—can easily get overexcited in this type of game and this is when accidents can happen. Keep your pet's training sessions short, but repeat the lesson frequently.

1 It is vital to have your cat's full attention when teaching this trick, so choose a quiet spot in the home where there won't be any distractions.

2 Hold your palm up and wave your fingers gently and slowly near your kitten's face, as shown. If you do it too fast, she is likely to strike out harder with her paw than you want.

This kitten is already showing promise as a high-fiver. She has her paw raised, and is sitting on the ground in a nicely relaxed fashion.

**3** When the kitten puts up her paw, touch it gently with your outstretched palm. Don't forget to say "high five" and have a treat handy to give her afterward.

★ Your cat sits and looks closely at your raised hand

★ ★ Your pet raises her paw gently toward your hand

★ ★ ★ She makes contact as you say "high five"

**Right:** Not all cats understand the need to play gently! It may be better to wear a pair of nonwoolen gloves to avoid scratches, particularly at first.

**Below:** It's a big ask to get a cat and dog high-fiving each other as shown here, but nothing ventured, nothing gained!

### Gently does it!
Give your cat a clear understanding of what you want her to do, by repeating the words "high five" as she lifts her paw toward your outstretched palm. This should help to ensure that this movement is carried out gently.

# Top of the Class

**H**ow is your cat scoring when it comes to mastering the basics? Don't forget that you need to be patient. Progress may not always be as smooth as you would like, but keep going, and you'll get there in the end!

Record how your pet is progressing on the special star chart opposite. Put the number of stars scored in each quiz into the boxes. For "True or False" questions score three stars for the correct answer but none for the wrong one. Add up the total number of stars your cat has scored in this quiz—you will need this for the final score chart on pages 186–187, when you will be able to work out just how smart your pet is.

## Getting to Know Your Cat

## Teaching the Basics

**How did my cat score?**

★ Mostly 1 star = more work needed!

★ ★ Mostly 2 stars = your cat is becoming a brainiac

★ ★ ★ Mostly 3 stars = your cat is a gold star pupil

# A Part of the Family

# Here Kitty!

If you decide to allow your cat to go outside into the yard remember that, inevitably, he is likely to wander farther afield. You must be able to persuade him to come back to you when called. You need to start training him to return home from the very first time that you let him out.

## Routine

Let your cat out first thing in the morning before you feed him, so he will return for his breakfast. Do the same in the evening before he has his dinner.

Cats cannot be let out safely until they have completed their first full set of vaccinations.

1 Be prepared to use a little bribery to get the response that you want in this case. It is really important that you can persuade your cat to come back to you without difficulty.

2 Accompany your cat outside when you first let him out and take a favorite treat along with you. Try to time this outing for a moment when your pet is going to be hungry.

3 Stand quietly by the back door and let him wander around for a time. Then call him back—if there is some food on offer he should return quickly. Repeat this routine until he gets used to the idea.

★ Your cat starts exploring in the yard

★ ★ Your pet comes back when he sees and hears you

★ ★ ★ He returns even if out of sight when you call

**Above:** Always praise your cat when he comes back to you when you call him in the yard. Use his name to reinforce the message.

**Below:** Let your cat come to you rather than chasing after him, as this could create problems if he starts running away.

# Being Vocal

Cats use a surprisingly wide range of calls to communicate with each other, as well as with people and other animals. These calls range from plaintive meowing through to blood-curdling shrieks!

1 Meowing is the most common feline sound, usually made in the presence of people. It is a plea for attention. If ignored, the cries often become increasingly loud and frequent.

2 Cats meow most commonly when they want food, but they may also make this sound under very different circumstances, such as when they are in pain or frightened.

A young cat will meow if he feels lost. This is a particularly common response following a move to a new home.

### Talking Siamese

Some cats are much more vocal than others, notably those with an extrovert personality. Typically members of the Siamese and Oriental breeds are like this. They are very demonstrative by nature and will engage in a distinctive dialogue with their owners by meowing.

3 When two cats are squaring up for a fight, they will start shrieking at one another in shrill tones. These calls become louder and increasingly intimidating if neither backs down.

**Below:** If frightened, as revealed here by the fearful body language, the "meow" call note is transformed into a throaty growl.

**Above:** It is quite common for a cat that is meowing to be purring at the same time, because he is happy and content.

**True or False?**
Cats can make over 100 different types of vocal sound, while a dog can only make about ten.

☐ True
☐ False

# Purr-fection

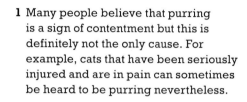

**P**urring is a puzzle. It is a very distinctive sound that tends to be associated with smaller members of the cat family and is a distinguishing characteristic that sets them apart from the so-called "big cats," like lions, that roar.

1 Many people believe that purring is a sign of contentment but this is definitely not the only cause. For example, cats that have been seriously injured and are in pain can sometimes be heard to be purring nevertheless.

2 It used to be thought that cats produce the purring sound as a vocalization but scientists are still uncertain exactly how cats manage to purr. They can even purr and meow simultaneously with their mouth closed!

3 A recent theory about the function of purring relates to the frequency of the sound, which is around 25Hz. Some scientists believe that it may be linked to healing, acting somehow to prevent the loss of bone density during long periods of rest or inaction.

Stroking a cat when he is relaxed like this, with his eyes closed, almost inevitably evokes a purring response. Some cats will naturally purr louder than others when being stroked.

★ Your cat starts purring when you stroke him

★ ★ Your pet rubs against you while purring, wanting food

★ ★ ★ He sits with you and purrs contentedly

**Above:** A mother cat will often purr in the presence of her kittens and they are also able to purr from an early age. This reciprocal sound may help them to communicate.

**Below:** There is no difference between the purring of ordinary street cats and purebred individuals. Some cats use purring as a means of seeking food and attention from their owners.

### Good health

Some alternative health practitioners believe that a cat's purring can promote good health in people. They believe that listening to purring can lower blood pressure and help wounds heal.

# Kids and Cats

Regular games with your kitten add a new element of fun to family life, but you must establish a few ground rules at the outset or it may all end in tears! Always keep an eye open when kids and cats play together, and teach children the warning signs.

1 Cats and kittens should always be treated gently and must not be teased, otherwise they may strike out. Quite apart from their needle-like teeth, they can easily inflict painful scratches with their sharp claws.

2 A kitten may pounce on your hand if you suddenly try to take a toy away, as he will think this is part of the game. So always move slowly and don't snatch away a toy suddenly.

3 Kittens love chasing after balls, and your pet will soon get into a routine of patting the ball to you in order for you to roll it back again. Kittens also like leaping on a length of string or ribbon being pulled across the floor.

Don't use hands as playthings! Kittens should not be inadvertently encouraged to pounce on them. Play gently at all times, and your kitten should learn to act with similar gentleness.

**Left:** Learning how to hold a cat securely is vital to avoid being scratched. Support the cat's body, but do not grip him too tightly.

**Above:** Cats can be very gentle but, just like children, young individuals can sometimes become overexcited and boisterous when they are playing with you.

**Fair play**

Cats can sometimes show frustration, if they do not get the ball to play with, for example, and they may react by lashing out with their claws. If a cat starts wagging his tail determinedly from side to side, this is a warning sign that he is getting angry.

★ Your cat is easy to pick up and enjoys being held

★ ★ Your cat plays with a range of different toys

★ ★ ★ He knows how to play simple games with you

# Close Companions

**C**ontrary to the popular cartoon image, cats and dogs will live together quite harmoniously in one household. In fact, they can form strong bonds and may play together regularly.

1 If you bring home a kitten at around the same time that you acquire a puppy, they will usually grow up together without any problems at all.

2 If you already have a dog and want to introduce a kitten into your home, don't force them together. Keep an eye on them when they are together, and keep them apart when you are out.

3 Quite often the cat turns out to be the dominant individual. However, some older cats refuse to engage with a dog living in the same household and habitually keep their distance.

This puppy and kitten grew up in the same home and ended up becoming the best of friends. However, the puppy has already learned caution, and he doesn't try to separate the kitten from the ball of wool.

★ Your cat is watchful of your dog but doesn't hide from him

★ ★ He interacts positively and plays with your dog

★ ★ ★ They both curl up and sleep happily together

**Left** Another happy couple, but do beware of dogs that instinctively chase cats, such as Greyhounds. At some point they will do so—and they are fast enough to catch them!

**Quick response**
A dog sharing a home with a cat will often rush out barking loudly to drive away any cat that threatens his companion. The dog will also recognize the cat's distinctive distress calls.

**Below:** Here are the same dog and cat shown in the picture opposite, after they have grown up! It is fascinating how they get on so well together.

# Furry Friends

If you watch lots of YouTube videos, you'll probably get the impression that cats get on with a whole menagerie of furry friends. Unfortunately, this is seriously (and even fatally) misleading. Why? Simply because most small furry pets are rodents or rabbits, which are a cat's natural prey.

1 Kittens tend not to have such highly developed hunting instincts as adult cats and so they are less inclined to attack and injure small pets, such as hamsters and rabbits.

2 Some cats have stronger hunting instincts than others, which puts small pets at greater risk. The most tolerant of the various breeds are Ragdolls, which are not inclined to hunt.

3 The greatest risk period is when a small pet starts to move, so triggering the cat's chase instinct. Hunting is a natural reaction which your pet is unlikely to be able to control.

It's not just your cat that you need to worry about. Being close to a potential predator could be very stressful for the rabbit.

★ Your cat chases after small pets

★ ★ He ignores small pets in the house

★ ★ ★ He peacefully sleeps close to a house rabbit

**Above:** A cat can find it very frustrating if a small pet lives in a cage just out of reach. He may cling to the cage and try to pull it over.

**Right:** A kitten encounters a guinea pig. Cats can find the noises uttered by these cute New World rodents particularly fascinating.

**Sharing the home**
Some rabbits show a strong reaction to the introduction of a kitten to their household, chasing the newcomer away. The chances are that a large rabbit will fare better in a house alongside a kitten.

# Got You!

Some cats develop an annoying habit of surprising and even scaring their owners by jumping out unexpectedly at them. This behavior tends to be a playful reenactment of the cat's natural hunting instinct.

1 It is not just unsuspecting people who may fall victim to a surprise feline ambush of this type. It is quite common behavior when a dog is living alongside the cat as well.

2 Your cat may select a favored hiding place in the home to leap out from, such as behind a chest or couch. In the yard, any bush may be used as cover from which to launch an attack.

3 The difference between encounters of this type and ambushing prey for real is that your cat will keep his claws in their sheaths when he pounces on you.

Some cats are not entirely trustworthy when you are making a fuss of them, and they may suddenly grab at you with their claws.

★ Your cat grabs you with his claws

★ ★ He grabs you but doesn't scratch you

★ ★ ★ Your pet quickly lets go of you

**Left:** Your cat may be watching you intently without your knowledge, and planning to ambush you as you walk through the door.

**Why do it?**
Aside from sharpening their reflexes, cats will pounce like this in order to draw a positive response from their owner.

Bengal cats are more inclined to use their claws than many other breeds, as their descent from ancestral wild cats only occurred relatively recently.

# Catnaps

**C**ats will often choose to sleep in the strangest of places around the home and sometimes their choice can even be slightly hazardous. Although you may provide your pet with his own bed, he may have other ideas about where he wants to sleep.

1 Warmth and comfort are not the only criteria governing choice of sleeping place. Your pet will also want to choose a quiet spot in which he feels safe and secure.

### Beware!

Cats can sometimes end up sleeping in very unusual and potentially dangerous places, even curling up under the open hood of a car. In the home they often choose to sleep under beds or in open drawers. If you have building work going on in your home, for example, your cat may vanish into a space under the floorboards for a snooze if he gets half a chance.

2 The Maine Coon breed, whose ancestors were often recruited as ships' cats, has a reputation for sleeping in the smallest and most unlikely places around the home.

3 Cats may change their resting spot according to the season, lying in an area that gets the sun in summer and retreating next to a radiator in winter.

★ Your cat finds odd places around the home to sleep

★ ★ He uses his appointed bed most of the time

★ ★ ★ He falls asleep next to you on a chair or couch

**Above:** Kittens need to sleep longer than adult cats. They often simply "flake out" and fall asleep after playing or eating.

**Right:** Their smaller size renders kittens more sensitive to the cold as their coats are generally shorter than those of adults.

# Hitching a Ride

If you have read the true story of Bob, a rescued street cat *(A Street Cat Named Bob)*, then you will already be familiar with a cat that would happily perch on his adopted owner's shoulders.

★ Your cat likes to sit curled up in your arms

★ ★ Your pet climbs up to stand on your shoulders

★ ★ ★ He settles quietly on your shoulders for a rest

1 Although it looks like a cute trick, remember that this is not a particularly comfortable position, either for you or for your cat. You are quite likely to be scratched by your pet.

2 The problem is that your cat needs to feel secure on your shoulders, otherwise he is likely to dig his claws into you as he struggles for balance.

3 When teaching your cat to sit on your shoulders, start by sitting down on an armchair so your pet can easily get on or off your body.

The risk of being scratched is greatest with kittens, simply because they are skittish and have very sharp points to their claws.

Don't wander around with your cat at first as he needs to learn to balance on your shoulders.

## Balancing act!

The first step is to train your cat so he is happy balancing on your shoulders while you are sitting still. Then you can try standing up and walking around.

# Golden Oldie

**W**hy is it that people only seem to want kittens when they are looking for a pet cat? Many older cats lie languishing in rescue centers in need of homes and, generally, they are easier to look after than kittens!

1 Most mature cats that have been rescued for whatever reason may already have been neutered, house-trained, and are used to home life.

2 Staff should be able to match you with a new pet well suited to your circumstances. Moreover, these cats will generally have had all the necessary health checks.

3 Since most cats now live into their mid- or late teens, choosing an older cat still means you can enjoy many happy years together with your pet.

★ An older cat approaches you confidently

★ ★ He is happy to be picked up and petted

★ ★ ★ He settles into your home without a hitch

**Below:** It may take time to win the confidence of an older cat. It depends largely on his past experiences.

**Above:** The placid nature of older cats, compared with kittens, means they often settle into a new home more quickly.

### Staying in

Remember that older cats, like kittens, must be kept indoors for several weeks before being allowed out, to stop them running away.

# Cat People

A confident cat that is settled in his home surroundings is usually quite happy to meet visitors. Unfortunately, not everyone likes cats and some people are allergic to them, so you must bear this in mind if you have guests in the house.

1 If your children's friends visit, they may not be used to cats. Supervise them to ensure they do not play roughly with your pet.

2 Should a friend bring a dog, move your cat to a safe haven in the home in advance, so he will not be distressed by this intrusion.

3 Don't force your cat to meet visitors, but give him the opportunity by bringing him in the room with you—although he may just walk out again!

Long-coated cats are most likely to cause allergic reactions in some people, which often result in repeated sneezing and runny eyes.

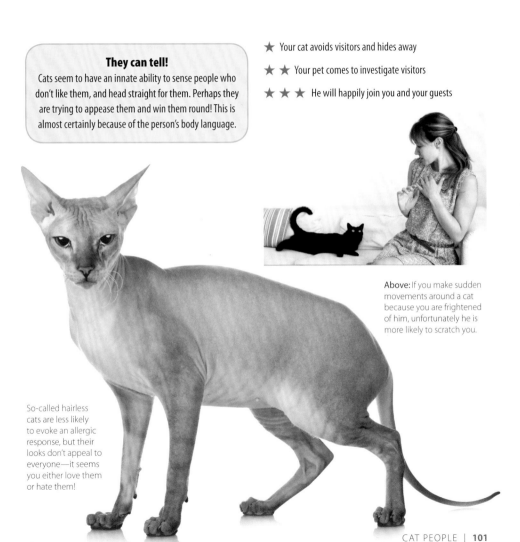

**They can tell!**
Cats seem to have an innate ability to sense people who don't like them, and head straight for them. Perhaps they are trying to appease them and win them round! This is almost certainly because of the person's body language.

★ Your cat avoids visitors and hides away

★ ★ Your pet comes to investigate visitors

★ ★ ★ He will happily join you and your guests

**Above:** If you make sudden movements around a cat because you are frightened of him, unfortunately he is more likely to scratch you.

So-called hairless cats are less likely to evoke an allergic response, but their looks don't appeal to everyone—it seems you either love them or hate them!

# Family Fun

**H**ere's your chance to see how your pet is progressing overall in terms of becoming a fully committed family member, joining in with different activities, and settling in happily alongside even the youngest members of the family.

Record how your pet is progressing on the special star chart opposite. Put the number of stars scored in each quiz into the boxes. For "True or False" questions score three stars for the correct answer but none for the wrong one. Add up the total number of stars your cat has scored in this quiz—you will need this for the final score chart on pages 186–187, when you will be able to work out just how smart your pet is.

Supervise young children and cats when they are together—cats can lash out if annoyed. Also keep them out of bedrooms where babies are sleeping.

# A Part of the Family

Pages 80–81
## Here Kitty!
☐ Number of stars

Pages 82–83
## Being Vocal
☐ True

Pages 84–85
## Purr-fection
☐ Number of stars

Pages 86–87
## Kids and Cats
☐ Number of stars

Pages 88–89
## Close Companions
☐ Number of stars

Pages 90–91
## Furry Friends
☐ Number of stars

Pages 92–93
## Got You!
☐ Number of stars

Pages 94–95
## Catnaps
☐ Number of stars

Pages 96–97
## Hitching a Ride
☐ Number of stars

Pages 98–99
## Golden Oldie
☐ Number of stars

Pages 100–101
## Cat People
☐ Number of stars

**How did my cat score?**

★ Mostly 1 star = more work needed!

★ ★ Mostly 2 stars = your cat is becoming a brainiac

★ ★ ★ Mostly 3 stars = your cat is a gold star pupil

Everyone should play a part in looking after the family cat. You don't want a kitten to only allow one member of the family to handle him when he grows up.

# Life
# Indoors

# Perfect Environment

**M**ore cats than ever are being kept permanently inside the home these days, because of fears about their safety outdoors. Owners are naturally worried about the potential dangers of busy roads and the risk of injury and infection when rival cats fight one another.

1 You will have to adapt your home if your cat is to spend all her time indoors. A suitable space must be designed to provide both physical and mental stimulation for your pet.

2 While you need to create a suitable floor area for your pet, remember that cats like to climb and so make adequate provision for her to do this too.

3 Places to hide and sleep need to be incorporated into the overall layout as well. In fact, specialist companies can be found that construct indoor play areas specifically for cats.

Even in the best planned homes, the active nature of kittens means that sometimes items get broken.

★ Your cat investigates her indoor play area

★ ★ Your pet willingly starts to use this part of the home

★ ★ ★ She exercises and plays regularly here

Kittens will soon feel secure in the home, curling up to sleep when they are tired, but they must be kept occupied when they are awake and active.

**Variety matters!**
Don't be afraid to move items around in your cat's area of the home to make things more interesting for her. Offer new toys for her to play with regularly as well.

Cats often appear calm and laid-back for long periods but they can also have frantic periods of activity, which you will need to plan for indoors.

# House Cat

Not everyone would choose to have a cat living permanently indoors, but sometimes it is unavoidable. Certainly cats can adapt very well to this type of domestic environment.

1 The key thing to ensure is that your pet gets enough exercise and won't become bored, as this is when behavioral problems may arise.

2 If this kind of domestic setup is to work, it also helps if you regularly stay at home for much of the day so that you can supervise your cat and ensure she is getting plenty of exercise and stimulation.

Having your cat indoors with you at all times gives you a greater opportunity to bond with your pet and to monitor her health and well-being.

**3** It is important to keep your cat's vaccinations up to date, particularly if your pet needs to go to a cattery unexpectedly for any reason.

**True or False?**
According to research in North America, about 80% of domestic cats live permanently indoors.

☐ True
☐ False

**Left:** Kittens can sometimes be destructive, so encourage your pet to use a scratching post away from furniture. Use blankets to protect upholstery and chairs.

**Above:** Cats gazing out of a window may make an odd chattering noise with their teeth. It usually happens when the glass panes stop them from chasing birds.

**Safety check**
By living permanently indoors, your pet will enjoy better protection from the various infections that can be spread by bites from other cats. The most significant of these is Feline Immunodeficiency Virus (FIV), for which there is currently no vaccine or cure available.

# Home Entertainment

**M**ake sure that you play regularly with your cat indoors, so that she remains fit and active. Cats soon start to make up their own games to play too.

2 Make sure that all such items are safe, and that pieces will not break off that might cause your pet to choke or create an intestinal obstruction.

1 Utilize the main play area that you have provided for your cat, adding plenty of toys that she can access, both down on the ground and suspended above it.

3 Leave lightweight balls on the floor that your cat can knock around with her paws. She will chase after them and pounce, as if hunting.

A toy of this type is very flexible and can be used as a lure, either dangled in the air or dragged along the floor for your pet to chase.

★ Your cat is keen to play with different toys

★ ★ Your pet follows your lead if you start a game

★ ★ ★ She starts playing games of her own accord

### Special diet
Keep your cat active indoors to prevent her from putting on weight and becoming obese. Swap your pet's food for a special indoor formulation. This will help to address health issues such as the drier air in your home, which may impact on the condition of your cat's skin.

**Above:** Cats often have favorite toys. Toys that make a noise can grab their attention, like this ball that contains a bell.

**Below:** Tunnels appeal to cats, as they like to explore inside confined spaces. They may even choose to sleep here as well.

# Boredom Busters

**P**laying is central to the lives of cats. Watch the way that a kitten will chase leaves blowing around outdoors for proof of that. Given plenty of opportunity to play, cats are much less likely to develop behavioral problems.

1 There are all sorts of different toys available from pet stores and online. Cats do differ in their preferences though, so you will need to experiment to find your pet's favorite plaything.

2 Toys that encourage cats to run and pounce are a good choice, as these encourage activity. Cats are naturally active in short bursts, so these complement their lifestyle.

3 Leave toys lying around for your cat at all times. This will allow her to play by herself if you are not around, and so helps to prevent boredom.

Some toys, such as this spring-loaded punchball and scratching area, are ideal to keep a cat occupied on its own when you are out.

**Left:** Climbing frames offer the opportunity for both climbing and scratching. This helps to keep your soft furnishings in one piece! They can also provide a handy sleeping platform too.

### Toys keep cats out of mischief

As hunters, cats are naturally alert and curious, possessing finely tuned senses. This allows them to locate the small creatures that normally form their diet in the wild. Play provides a means to keep them happily occupied in the home, so they are less inclined to cause damage there. Kittens instinctively play for longer than adult cats.

★ Your cat is interested in playing with toys

★ ★ She responds to playing games with you

★ ★ ★ Your cat initiates a game by herself

**Right:** Cats are naturally inquisitive and like to explore hidden corners, such as inside a paper bag. These can also double up as favorite places to lay an ambush.

# Up to Scratch!

A cat's claws make a huge contribution to its quality of life in a number of different ways. They help cats to climb with agility, to catch their prey, to groom themselves, and to defend themselves against attack.

1 Cats—whether living indoors or out—depend on their claws. In some parts of the world it is permitted for cats to have their claws surgically removed but this practice is emphatically not to be recommended.

2 The loss of her claws can seriously handicap your pet if she should manage to slip out into the wider world outside your home. It also handicaps her indoors.

3 You should encourage your cat to use her claws on a scratching post indoors to prevent damage to the furniture or furnishings. Having her claws extracted is draconian.

This toy is ideal for keeping a cat's claws healthy. The central part is a scratching post while the hanging ball is a tempting target to attack.

★ Your cat sniffs at the scratching post and is interested in it

★ ★ She starts to use it when you show her gently how to

★ ★ ★ She scratches on the post regularly

**Below:** If space is limited, you can make your own custom scratching post with a piece of thick carpet. Make sure that this is set at the correct height for your pet.

The benefit of special cat furniture is clearly demonstrated here. The marks of the cats' claws are evident under the platform.

### Trim tips
The tips of a house cat's claws can easily become overgrown because she is not wearing them down outdoors. Your vet will be able to trim them.

# Play Rewards

**P**laying with your cat regularly is a great way to develop your individual fun routines together. Remember to praise your cat frequently and reward her with an occasional treat.

1 The type of games you play will depend on the sort of toys that your cat is given. Keep things calm. You don't want to cause your pet to play in an aggressive way by getting her overexcited.

Keep control of a play session by slowing down the pace now and then and giving a treat, especially with a young excitable cat.

2 Some toys are more likely to trigger an aggressive response than others, particularly any that your pet chases after as if they were prey. This game can lead to you being scratched.

3 If your cat rolls over, kicking out fiercely with her hind legs while trying to bite the toy, don't get involved. Allow this mad moment to subside.

★ Your cat chases a toy around while playing

★ ★ She pauses when you offer her a favorite treat

★ ★ ★ She lets you stroke her as part of the game plan

**Below:** Treats are one way of calming down an overexcited cat; praising your pet and making a fuss of her is another tactic to slow the game down.

**Above:** Kittens are naturally more playful than adults as they learn to hone their innate hunting reflexes.

### Tasty treats!

Various types of treats are available and cats often develop their personal preference. Feel free to experiment and find one that appeals to your cat. Treats tend to be fish- or meat-based but they are not balanced nutritionally, as commercial cat food is.

# Sniffing Out a Treat

The way a cat frenetically chases after something becomes very apparent when you play together. What is less obvious is how important your cat's sense of smell is to her hunting games.

**2** The external surface of a cat's nostrils has quite a leathery texture which helps to protect this part of the nose from injury. It may also help with the detection of certain scents.

**1** Cats rely heavily on their sense of smell. They use it to detect if other cats have recently passed through their territory, to locate potential mates, and the whereabouts of prey.

**3** A cat's sense of smell is not as acute as that of a dog, but there still can be as many as 80 million scent receptors in her nose.

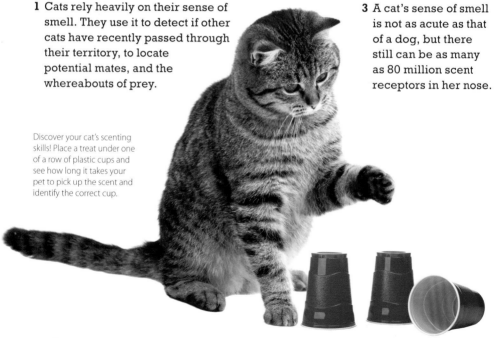

Discover your cat's scenting skills! Place a treat under one of a row of plastic cups and see how long it takes your pet to pick up the scent and identify the correct cup.

★ Your cat sniffs at a treat ball and pushes it around

★ ★ She paws at the treat ball until she gets a reward

★ ★ ★ Your pet can find a hidden treat ball by its scent

Your cat will soon learn to recognize your personal scent and will be able to distinguish you from other household members just by sense of smell.

Treat balls are popular with cats of all ages. As they discover the scent and push the ball around with their paws, they are rewarded with tasty tidbits falling out.

### First sense
It's thought that a cat's sense of smell is about 14 times more powerful than our own. Smell is the first sense that newly born kittens use before their eyes open.

# Hunt for the Treat

To make the hunt for a hidden treat as successful as possible, you must first discover your cat's favorite flavor. This preference can then be used to test her scenting abilities.

1 Obtain several different types of treat, lay these out on the ground, and then note which one your cat approaches to eat first. This is probably the favorite flavor.

2 Discard any at which she turned up her nose. Place four of the more popular varieties around the room. They do not necessarily all have to be hidden, but space them apart.

3 Let your cat in and see how long it takes her to find the hidden treats and watch to see which she eats first. By repeating the hunt at intervals, you will confirm which she loves most.

Cats are less likely to be interested in playing this game if they have just eaten a meal.

★ Your cat is interested in seeking out the reward

★ ★ She sniffs around methodically to locate the treats

★ ★ ★ She finds a treat both on the ground and up high

**Left:** Cats should only be given treats sparingly and in small quantities, especially those that are calorie-rich. This is to prevent unwanted weight gain.

**Below:** If you are happy to let your cat outdoors, you can play a game of hide 'n' seek with her by concealing treats in your yard for her to seek out.

### No trick, a treat

Bear in mind the aim of the game is to encourage your cat simply to seek out the treats, rather than providing them as a substitute meal. The emphasis of the game should be on your cat having fun, and not on her receiving a string of big rewards!

# Cat Burglar

**C**ats are very skillful at seeking out tasty items of food that appeal to them, relying on their sense of smell for this purpose. Unfortunately they don't just follow their noses on your property. They are quite likely to wander farther afield, and may even start sniffing around your neighbors' barbecues.

**1** The smell of meat or fish cooking will often attract your cat to the kitchen. This behavior will be reinforced if your pet is given a piece to taste.

Cats eagerly seek out any freshly cooked meat. Take care not to give your pet any potentially harmful items, such as chili-flavoured sausages or spicy meats. Always allow tidbits to cool down first.

**2** If you cook meals that give off tasty aromas at regular times each day, your cat will spot this opportunity for food and start to show up in anticipation.

**3** Always dampen down barbecues when you have finished cooking to eliminate the risk of a cat burning itself when following the scent.

**Left:** You don't want to own a cat that is a scavenging nuisance around the kitchen. It's not hygienic and the food itself may harm the cat.

> **Beware!**
> Never leave your cat alone with meat or fish cooking on a stove indoors or a barbecue grill. They may burn themselves badly if they leap up to reach the food.

★ Your cat shows interest when you are cooking

★ ★ She displays "cupboard love," weaving around your legs

★ ★ ★ She routinely appears whenever you cook meat

**Above:** Preserved meats such as ham are unhealthy for cats, because of their high salt content.

# Cat-nipped!

Catnip produces attractive small purplish flowers and is easy to grow.

It has been known for centuries that cats react quite strongly to various plants. The Roman writer Pliny observed that cats do not like to enter areas planted with the herb rue. Conversely, catnip or catmint appeals to many cats, and is often grown specially for them.

1 Catnip (*Nepeta cataria*) grows naturally in southern areas of Europe, across the Middle East, and Asia, as far as China. It has been introduced to the wild in many other parts of the world.

2 The plant contains an active ingredient called nepetalactone. This can be extracted commercially and is used in the manufacture of cats' toys. It makes them more appealing, although ultimately the scent dissipates.

3 Studies suggest that only around a third of cats react to this plant. Kittens under 12 weeks show no interest in it. It appears to be an inherited trait that only attracts some cats to catnip.

Cats living indoors will enjoy eating fresh grass and you can buy complete cultivation kits for this purpose. Catnip can also be easily grown in a pot on a windowsill.

**Above:** Cats react to catnip by rubbing themselves against the plant, which is a member of the mint family. The scent has a limited lifetime, lasting anywhere from five to 15 minutes.

**Below:** Some cats become excited when exposed to catnip; others become sleepy and relaxed.

# Opening Doors

**S**ome cats master the skill of opening doors by watching their owners do it. Of course, this makes it harder to confine a pet that has learned this trick. Nevertheless, it also provides dramatic proof of the remarkable way that cats can learn by observation.

1 Luckily, in most cases, the door handle to a room will not be the right shape for easy manipulation by a cat, or it will be out of her immediate reach.

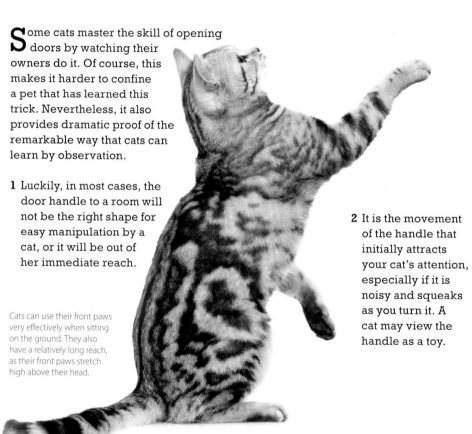

Cats can use their front paws very effectively when sitting on the ground. They also have a relatively long reach, as their front paws stretch high above their head.

2 It is the movement of the handle that initially attracts your cat's attention, especially if it is noisy and squeaks as you turn it. A cat may view the handle as a toy.

★ Your cat watches intently when you turn a door handle

★ ★ She reaches up to play with a handle as it moves

★ ★ ★ She has learned to use the handle to open a door

3 Although it is a clever skill, problems can arise once your cat has worked out how to open doors by using her weight to turn handles. You've now got a potential escape artist on your hands.

**Left:** Some cats, such as Siamese, use their front paws to grab items above their heads, grasping tightly onto their target.

**Above:** Your cat may even jump up and seize a door handle, using her hindquarters to help to support her weight.

# Roll Over

A cat that is relaxed and who wants attention from her owner will often roll over on her back to enjoy having her tummy gently stroked. A little praise will help to encourage this behavior.

1 Cats most commonly stretch out when they are warm and comfortable. A cat will frequently sunbathe in this way, lying full length on the warm ground.

A cat will often lie on her side when she is resting and will enjoy being made a fuss of by her owner.

2 When approached, your pet may then roll over onto her back for you to stroke her. Do so cautiously as some cats can react rather unpredictably when petted in this position.

3 This is because a cat can often feel vulnerable lying on her back. She may paw at you, or even attempt to bite your hand if she feels that she is being held down and is unable to escape.

★ Your cat stretches out when warm and comfortable

★ ★ She rolls onto her back when you approach

★ ★ ★ She lies on her back while you stroke her tummy

**Right:** Kittens often enjoy being stroked on their underside. But they may suddenly decide that it's game over and make strenuous attempts to get up.

**Below:** What begins as pleasant bonding with your pet can sometimes develop into a sharp encounter with claws exposed, so watch out.

### Work out!
Rolling around on the ground emphasizes the flexibility of a cat's body as she stretches and tones her muscles. She may even wriggle along the ground a little on her back.

# Play Station

**P**urpose-made play stations provide a great way to keep your pet occupied indoors. These multipurpose units are suitable for cats of all ages and are ideal for fun and games around the home.

1 With many designs available, choose a play station featuring activities that are likely to appeal to your cat. A scratching post is a useful addition as it will deflect your pet's attention away from your furniture.

2 Unpack the play station with your cat in the room, so she will come over and investigate. Introduce the toys, drawing her attention to each of them.

3 Encourage your pet to use the play station over the next few days. A quick spray with catnip every now and then may arouse your cat's interest in it.

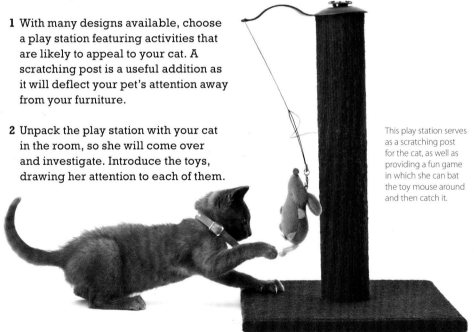

This play station serves as a scratching post for the cat, as well as providing a fun game in which she can bat the toy mouse around and then catch it.

★ Your cat sniffs at the play station

★ ★ She starts to use the equipment

★ ★ ★ She exercises on the play station regularly

The toys that are part of a play station keep cats occupied, even when you are out.

**A home gym!**
It is a good idea to set aside an area of your home for your cat, which is free from obstructions and where your pet can run around and play to her heart's content.

Play stations are sometimes more accurately described as cat furniture. This unit incorporates toys, scratching post and tunnels for a cat to explore and have fun in.

# Hide and Seek

Cats love to explore quietly on their own, often prowling intently around the home and climbing into open drawers or cupboards. This behavior reflects their inherent curiosity, which is apparent from kittenhood.

1 Cats seem to find empty cardboard boxes particularly fascinating. They will spend hours clambering into them and settling themselves down to rest.

2 A cardboard box lying on its side, with the flaps partially closed, is an irresistible target for most cats. They will use their front paws to open up the flaps and then crawl in.

**Left:** From an early age cats are curious about their surroundings and are keen to explore everything.

**Above:** If your cat disappears from the home, she may be hiding in a secluded spot, such as a box.

**3** Your cat will sniff cautiously before entering the box to make sure it is safe. She may then just decide to curl up and fall asleep inside, assuming the box is big enough. Sweet dreams!

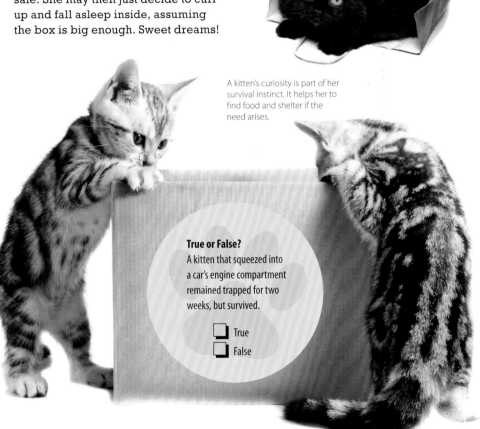

A kitten's curiosity is part of her survival instinct. It helps her to find food and shelter if the need arises.

**True or False?**
A kitten that squeezed into a car's engine compartment remained trapped for two weeks, but survived.

☐ True
☐ False

# Social Climber

Cats are well-equipped to climb—they are strong, supple, and have great balance. Outdoors, cats can find trees to climb but those living inside need other options.

1 Suitable climbing frames for cats can be made quite easily if you have access to pieces of sawn timber or thick branches. The key is to ensure that the base of the frame is secure, so that it will not topple over with your cat's weight as she climbs it.

2 Cats will often climb trees or fences to give them better a vantage point from which to view their territory. Keen hunters have great eyesight.

3 Another reason that cats will climb is to escape danger. Dogs and foxes cannot pursue them up trees.

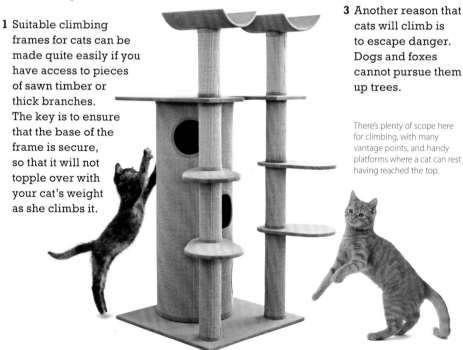

There's plenty of scope here for climbing, with many vantage points, and handy platforms where a cat can rest having reached the top.

- ★ Your cat begins climbing in the home or garden
- ★ ★ She rests happily on a platform off the ground
- ★ ★ ★ Your pet comes down without a problem

Young cats are especially likely to climb shelving in the home. Check that units are firmly supported, securing them to the wall with screws if necessary.

**More experience needed!**

Kittens are not as adept at climbing as adult cats and so they are more likely to get into difficulties.

# Water Splash

I't's a myth that cats totally dislike water. In fact, some wild cats like jaguars will actually hunt in water. What domestic cats often resent, however, is being bathed or sprayed with a jet of water.

1 Domestic cats have a dense coat that prevents them becoming saturated and chilled when it rains or snows outdoor. The outer layer of their fur is water-resistant.

Cats are not as averse to water as is popularly believed. In the wild they often hunt in water to catch fish.

You don't normally need to bathe a cat unless the coat is badly soiled. Some cats (unlike this individual) will react aggressively if you do try to bathe them.

**2** As a result, cats can simply shake their coat to remove unwanted water droplets. This applies to shorthaired and also long-coated cats, whose fur is denser and more profuse in winter.

**3** If your cat's coat does become wet for any reason, dry it as much as possible with a paper towel. Never use a hairdryer—this will only upset your pet.

★ Your cat struggles if her coat gets wet

★ ★ Your pet will sometimes sit in a bowl of water

★ ★ ★ She lets you bathe her without protest

Cats can become fascinated by running taps, watching the water fall. They will sometimes drink from the flow using their tongue rather like a curved ladle.

**Keeping it dry**
If you ever need to bathe your cat, avoid wetting her head if at all possible because many cats find this upsetting.

# Feline Movie Star

There is very distinguished roll call of feline movie stars. Now your cat could join them! Thanks to the Internet, your pet could be grabbing an international audience of millions before you know it.

1 Is your cat a diva? Just like all good movie stars, your pet must be willing to wait around so that the film sequences are just right. She also has to be in the right mood to perform.

2 Set-dressing is important, so plan your movie shoot in advance. Pay attention to the background you choose so that it doesn't distract from your pet's starring performance.

3 Make sure that the movie set is quiet and there are no obvious diversions that will distract your cat from the job in hand. Having a director's assistant on set is likely to be very helpful.

You want your cat to be as relaxed as possible and well rehearsed for her starring role. It is generally easier to get good still photos rather than video.

★ Your cat can stay sitting still

★ ★ She poses with props and lights

★ ★ ★ She poses happily with a dog

**Grab the shot!**
It is vital to have everything ready to roll so that you don't have to adjust the camera or lights at the critical moment. There may not be a second chance!

Don't put your pet at any risk of injury, however great the shot may look. You can always carry out some photo-editing later to improve on the end result.

# Home Sweet Home

Cats generally settle well in domestic surroundings and this helps to explain why they are such popular pets today. Whether you obtain a young kitten or an older cat from a rescue organization, both should adapt well to living with you in your home.

Record how your pet is progressing on the special star chart opposite. Put the number of stars scored in each quiz into the boxes. For "True or False" questions score three stars for the correct answer but none for the wrong one. Add up the total number of stars your cat has scored in this quiz—you will need this for the final score chart on the pages 186-187—and then you will be able to work out just how smart your pet really is!

## Life Indoors

### How did my cat score?

★ Mostly 1 star = more work needed!

★ ★ Mostly 2 stars = your cat is becoming a brainiac

★ ★ ★ Mostly 3 stars = your cat is a gold star pupil

# Establishing Territory

# Why Cats Roam

All cats love to explore their environment, whether indoors or out. Obviously, when outside they are able to roam farther afield. Unneutered pets are the most likely to head off and roam, often in search of a mate.

1 A kitten venturing outside will need to establish himself in an area which is probably already occupied by other cats. This can lead to fighting between them and may result in your cat being forced even farther away from home.

2 If you adopt a stray, be sure to keep your new pet inside for several weeks, so that he comes to recognize his new place as a safe haven. His home comforts should reduce his urge to roam, especially if he is neutered.

3 Some cats, especially those with strong hunting instincts, will travel quite a long way, particularly during spring and summer when there are plenty of young rodents and birds around for them to catch.

Once your pet is well established in his home territory (and especially if he has been neutered), he will be less inclined to roam far from home in the normal course of daily life.

★ Your cat likes to venture outdoors

★ ★ Your pet will come back indoors when called

★ ★ ★ He remains close to home and doesn't stray

Cats that roam widely face a number of risks, particularly the possibility of being hit by vehicles. Other hazards include accidental poisoning, predators, and territorial disputes with other cats.

**Gone roaming**
Studies have shown that pet cats generally don't roam over an area much larger than 5 acres, whereas feral cats range over much wider territories. This is probably because feral cats need to travel farther to find reliable supplies of food.

# Getting Along

Cats living in urban areas coexist in much higher densities than they do in the wild, where numbers are largely influenced by the availability of food. Even after being neutered, some cats can remain quite aggressive by nature. These are the animals that may wind up with the reputation of being neighborhood bullies.

Snarling and hissing serve as part of the intimidation process. The white cat, with his ears flattened, is being threatened by the tabby.

1 Cats establish networks of invisible paths to cross each other's territories. Conflicts result when one comes into contact with another, and the intruder will not back down when challenged.

2 When confronted, cats go through a ritual of gestures that are intended to intimidate a rival before launching into a physical attack. If a fight breaks out, contact is generally brief.

**3** If your cat is about to get into a fight, you are likely to hear him before hostilities break out. The noise the cats make increases in intensity if neither animal is prepared to back down.

★ Your cat wanders around the neighborhood

★ ★ Your pet avoids getting into fights with other cats

★ ★ ★ He makes friends with the other cats he meets

**Left:** Some cats in a neighborhood may become friends, even if they live separately. Typically, they will have grown up together.

**Below:** Other individuals just do not get along, for no obvious reason. They may clash regularly whenever they encounter one another.

**Watch out!**
An abscess often results if your pet is bitten by another cat. The affected area swells up, feels hot, and your cat will seem off-color and will lose his appetite. Seek veterinary advice as a bite from another cat leaves unpleasant bacteria in the wound that will need treatment.

# Outfoxed

In many areas the urban fox population has grown significantly over recent years. This upsurge brings foxes increasingly into contact with cats, especially at night but also potentially during the daytime. Some owners fear that foxes could be a threat to their pets, particularly to elderly cats and kittens.

**1** Don't let your cat roam outdoors at night—not just because of foxes, but also because the risk of road accidents increases after dark.

**Foxy feline facts**

Studies have revealed that foxes represent little danger to most cats, rarely interacting with them, but older, less agile individuals and vulnerable kittens are most at risk.

**2** Cats are more agile than foxes, not least because they are able to climb very well, making it relatively easy for them to escape from a fox if they find themselves threatened.

Leaving food in your yard for urban foxes at night adds to the risk of conflict breaking out with cats.

**3** If a cat is unfortunate enough to be cornered by a fox, he is well-equipped to defend himself, with very sharp claws and a painful bite if provoked.

Remember that urban foxes can be active at any time during the day, and not just at night. Removing foxes from urban localities is not a long-term solution—the missing foxes will soon be replaced by others.

**True or False?**
Foxes habitually spread a variety of parasites to cats in the outdoor environment.

☐ True
☐ False

# Cat Cam

Modern technology means that it's now easier than ever to keep an eye on your cat, both while he is indoors or roaming in the yard. One of the main advantages is that you will be able to check your pet's safety via a remote camera even if you're away from home.

1 The majority of cats do not react to seeing their image on-screen, so although you can monitor them, they are almost certain to ignore you!

2 There are small cameras available that you can attach to your cat's collar. They allow you to see where your pet is going when outdoors.

3 Make sure a camera of this type is safely fitted, however. You don't want your cat to snag his collar and the attached camera on a branch as he stalks through the undergrowth.

A basic webcam is suitable to view your cat, but he may become intrigued by it and see it as a toy to pat around with his front paws.

★ Your cat runs away while you are filming him

★ ★ He is happy to lie around while he is being filmed

★ ★ ★ Your pet shows off in front of the camera

**Left:** It's great to have a record showing your pet growing up and changing as he becomes older. You can even use his photo as your online avatar.

**Below:** Tablets provide an easy and convenient way to record your cat's activities, whether in the form of video or still photos.

### On the record

Scientists have been using webcams to carry out serious studies of cat behavior. They are ideal for this purpose, as they record what cats are doing naturally without distracting them in the way that the presence of a human observer is likely to do.

# Leaving a Mark

You're sure to notice that your cat sometimes uses his front claws to scratch at items around him, both outside and inside the home. If your pet attacks furniture in this way, it will not be long before it is bearing the tell-tale marks, so it's best to sort out this problem behavior as soon as possible.

## CLAWS AND PAWS

1 As predators, cats naturally need to keep their claws sharp. They rely on their retractable claws to pin down their prey so they have an opportunity to inflict a fatal bite.

2 Cats also use scratching as a means of communication. By leaving claw marks in prominent positions, your pet is laying down a visual claim to his territory to deter other cats.

3 There is a hidden marker at work here as well, because cats have scent glands between their paws. As your pet scratches with his claws, he is also leaving his individual aroma on the object of his attention.

Cats scratch regularly at specific locations around their territory—often on trees or fence posts. These marks will remain obvious, even if the rain washes away their distinctive scent.

### Scratching post

Scratching indoors can damage furniture and furnishings, so invest in a scratching post. Teach your cat to use it by gently moving his paws and claws down the post.

★ Your cat scratches items both indoors and outside

★ ★ Your pet sometimes uses a scratching post

★ ★ ★ He always uses a purpose-made scratching post

**Left:** Although cats usually scratch close to the ground, supporting themselves on their hind legs, they can also use suitable horizontal surfaces as scratch pads.

**Below:** Cats sometimes appear to leave a claw embedded in the wood where they have scratched, but this is just the discarded outer casing of a claw and is quite normal.

# Top Cat

**K**eeping two cats together can lead to problems, unless they are kittens—ideally littermates—or are older cats that have grown up together. Trying to integrate two cats from different backgrounds into a household can be very difficult, particularly if one of them is already well established.

1 There is no guarantee how cats will react until they are together. They may ultimately get on, but each will be wary of the other at first.

2 Let them meet on their own terms. Never force them together by shutting them in the same room, as this could very easily result in a fight.

3 Be prepared for a breakdown in toilet training, particularly in the case of an established pet. This is because cats use their body waste as a means of marking their territory.

★ Your established cat hisses at the newcomer

★ ★ They live together in relative harmony

★ ★ ★ They have established a peaceful hierarchy

**Left:** Things may be calmer outside. Both these cats are clearly relatively relaxed, as can be seen from their body language. The black cat has slightly flattened his ears to show that his attentions are not aggressive.

**Below:** Not all apparent conflict between cats is actually serious fighting. Play-fighting can be seen between adult cats, as here, especially with kittens that have grown up together.

### Shy by nature

Always remember that cats, unlike dogs, are not instinctively social by nature. This means that keeping two or more cats together in a household can cause them stress. There is also a risk that one could be forced out by the other.

# Cat's Eye View

Cats benefit from being outdoors, if it is safe for them, as they will take more exercise and generally stay fitter in natural surroundings. Although you won't be able to confine your cat outdoors, you can design your yard to make it as safe as possible for your pet. Build a tall secure wooden fence, with a minimum height of 6½ feet, around the yard to discourage him from jumping out. Plants can also play their part in preventing your pet from straying. Good thick hedges form useful natural barriers. Don't forget to make the yard attractive with plants that cats often like, such as catmint and catnip.

> **Did you know?**
> Some cats seem to enjoy a natural "high" when they roll in and chew catnip leaves. The plant has been described as a nonaddictive, legal, recreational drug for cats!

**Left:** One cat rubbing against his companion is a sign that these two friends enjoy a definite bond. This behavior merges their scents indicating that they are happy to be in one another's company.

**Below:** Master of all he surveys . . . Your pet may find a favorite vantage point from where he can look out over his territory, and where he feels safe and secure.

**Left:** Cats sometimes fight over their territory. The more dominant individual is likely to guard his patch and keep others away from it.

# Play Time

Play is a very important activity for domestic cats, helping them to exercise their bodies and stay fit, as well as developing their coordination. Although kittens are naturally most inclined to engage in games, nearly all cats will enjoy playing them through their adult lives and well into old age.

1 You can choose from a wide range of toys for cats these days. It is often the movement of the toy that will most attract your cat's interest to it.

2 Cats often have their personal favorite toys and what appeals to one cat may not appeal equally to a companion. Be prepared to experiment to find what suits each individual best.

It is often harder to take a toy from a cat than it is from a dog. Cats are not so easily trained to give up their toys willingly.

★ Your cat sniffs and investigates a toy

★ ★ Your pet starts playing a game with you

★ ★ ★ He begins to play by himself with his toys

### Care with toys

Most cat toys sold today are quite safe for your pet, but be careful with items that are not designed specifically for this purpose. A small ball, for example, could cause your pet to choke if swallowed. Cats often make up their own games, using random household items.

**Left:** Cats generally like to join in games where they are encouraged to chase a toy or stick, as this is a natural part of their hunting instinct.

**Below:** Cats love things that move, like these bubbles glistening in the air. Your cat may happily start trying to catch them with his paws.

# Off the Ground

Cats can climb well, thanks to a combination of sharp front claws and powerful hind legs. These physical attributes mean they can clamber quickly up a tree or fence post with relative ease, and then walk confidently along a branch or the top of the fence.

1 Young cats tend to have more difficulty in climbing than older individuals, because they are not as strong, nor as experienced.

2 Cats soon establish their own special climbing trails. Your pet may routinely head along a particular fence to reach a low, secluded roof where he can laze quietly in the sun.

3 If startled, a cat will leap down from a fence with great agility. If frightened by a dog on the other side, for example, the cat uses the barrier to avoid a dangerous confrontation.

When walking along a narrow fence, a cat uses his tail to help him maintain his balance. He will move carefully, taking relatively short steps, and keeping his head down.

★ Your cat likes to climb up on fences in the yard

★ ★ Your pet walks easily along the top of a fence

★ ★ ★ He jumps nimbly from the fence to get down

### Paw-fect landing!

If a cat does tumble from a significant height, he will instinctively swivel his body round in midair to land on his feet, which act as efficient shock absorbers. It is rare for cats to suffer serious injury as a result of a fall.

Cats generally move quite slowly and deliberately off the ground, in order to minimize the risk of slipping. They use their front paws for feeling the way forward, and their stronger hind legs to support themselves.

# Bird's Eye View

**M**any cats enjoy climbing and some literally take this activity to a higher level by heading up into trees. They may short-circuit the climb by leaping onto the tree from a convenient fence or shed roof, and then continue to shin up through the branches.

Some breeds, like the Siamese seen here, are more natural climbers than others. They may even try to go straight up your drapes indoors!

1 When a cat scales a tree, he does so in a measured way, using his front legs and claws like a climber's crampons, powering himself up by his hindquarters.

2 Once he reaches a suitable horizontal branch, he may walk a distance along it, or simply rest there with his feet dangling down on either side.

3 Coming down is less elegant. A cat simply descends backward, until he gets close to the ground, and then he turns round and jumps down the final distance.

**True or False?**
A cat abandoned at a day old regularly goes rock climbing with her new owner.

☐ True
☐ False

It's often thought that cats can get stuck up trees but usually, if left alone in peace, a cat will find his own way safely back down again.

**Indoor activity**
Special cat climbing frames (often doubling up as scratching posts) are now available for you to buy for your pet. Many cats enjoy playing on these indoors, especially if they are not allowed outside.

# Jump To It!

**C**ats are great all-round athletes! They not only climb well, but they are naturally agile jumpers too. You may not notice this skill in the home that often, until the day perhaps when your pet mistimes a leap onto a shelf and knocks over a precious ornament.

1 A cat will jump out of the way if frightened, perhaps as the result of an unexpected encounter with a dog. Young cats typically leap farther than older individuals, as they are naturally more athletic.

2 Cats use their muscular hind legs to propel themselves forward and upward when they jump. They always land on their front feet first, which helps them to keep their balance.

A cat in midair is an impressive sight. As well as arcing forward like this, cats can also jump vertically, springing up on their hind legs to catch something high above them.

**3** A cat's keen eyesight is a very important tool when it comes to jumping. It allows your pet to assess the distance of a jump accurately, and so helps to guarantee a safe landing.

★ Your cat investigates a hoop and will walk through it

★ ★ With prompting, your pet will jump through it

★ ★ ★ He jumps through readily in both directions

It's fun to teach your cat to jump through a hoop. Keep it low at first and offer a treat to encourage your pet to spring through to the other side.

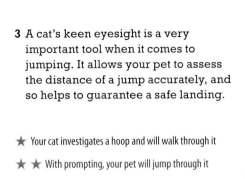

### For the high jump!
Domestic cats can jump distances of up to 6½ feet, but the African serval, a medium-sized wild cat with particularly long legs, is able to spring nearly 9 feet vertically upward from the ground.

# Skateboard Fun

There are some remarkable videos on YouTube showing cats apparently enjoying skateboarding. Don't be fooled! These individuals are the exception rather than the rule. After all, if all cats could skateboard with ease, who would want to watch them!

## COOL RUNNING KITTY

1 The first step is to get your cat used simply to sitting on the skateboard while it's stationary. He is likely to be uncooperative at first, but be patient.

2 Once your cat is happy sitting on the board, you can then start to push it along very slowly. Don't be surprised if your pet jumps off at first!

3 After a time your cat may be happy to stay on the board as it moves. Then you can start pulling the board along over longer distances.

This kitten seems quite comfortable sitting on a skateboard. It is easier if you start this with a young cat.

★ Your cat gets used to walking over the skateboard

★ ★ He sits happily on the stationary skateboard

★ ★ ★ He stays on the skateboard when you move it

### "Sick" tricks

Cats are generally not as responsive to training as dogs. This is largely because they are more independent by nature. Their inherent caution can also make training for a trick like this more difficult. Nevertheless, contrary to popular belief, with plenty of patience you can teach your pet to carry out a whole range of unusual activities.

You must be patient with your cat, and don't forget to give him a treat when he responds in the way that you want him to. This helps to reinforce the lesson that you are teaching.

# The Wider World

So here's your opportunity to assess how well your cat is progressing when out and about exploring the wider world around your home. How well has he settled into this unfamiliar environment? Roaming around outdoors will help to ensure that he gets plenty of exercise, lessening the risk of obesity, particularly for cats getting on in years. Your pet should also become more confident as a result.

Record how your cat is progressing on the star chart opposite. Put the number of stars scored in each quiz into the boxes. For "True or False" questions score three stars for the correct answer but none for the wrong one. Add up the total number of stars your cat has scored in this quiz—you will need this for the final score chart on pages 186–187, and then you will be able to work out just how smart your pet really is!

Kittens will generally prove to be more responsive to training than older cats, as they are able to learn more quickly when young.

## Establishing Territory

Cats rely on their instincts when exploring the world around them outdoors. Their instinctive caution helps to keep them out of danger when they are out and about.

**How did my cat score?**

★ Mostly 1 star = more work needed!

★ ★ Mostly 2 stars = your cat is becoming a brainiac

★ ★ ★ Mostly 3 stars = your cat is a gold star pupil

# Farther
# Afield

# Going Somewhere?

It's a cat owner's worst nightmare! All the preparations are made, the cases are packed, and you've made arrangements with the cattery—and then your pet disappears! Cats are remarkably sensitive to changes in their environment and adjustments to their daily routines. These can be enough to cause them to do a sudden vanishing act.

## PLAN AHEAD

1 Don't leave it to the last minute to take your cat to the cattery when you are going away. Arrange the drop-off for the day before, so there should be no last-minute panic.

2 Try to arrange the time of your drop-off appointment for first thing in the morning, so you can safely get your pet into the house the night before.

3 This should ensure a trouble-free trip to the cattery in the morning, although you may need to provide a litter tray for the overnight period.

Cats often become nervous when placed in a carrier, and they can become very noisy on the journey. This is because they start to associate traveling with what to them are unwelcome experiences, such as visiting the vet.

★ Your cat is reluctant to go into the carrier

★ ★ She goes into the carrier without a fuss

★ ★ ★ Your cat travels quietly in the carrier

Some cats get quite nervous when taken to the vet. Try to arrive slightly ahead of time, so that your pet can settle down after the journey.

# Walk on the Wild Side

Preventing your cat from hunting isn't easy, as this remains a strong instinct in many individuals. They can be cunning too, soon sidestepping any attempts on your part to warn potential targets of their presence. Sometimes, though, nature turns the tables and the hunter becomes the hunted! Playful kittens are especially at risk of being stung by wasps or bees.

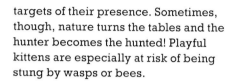

1 Try to persuade your cat to wear a safety collar and bell. She will probably resent this at first, and try to force the collar off with her paws.

2 Before long though, your pet should be running around quite happily. But now, when she stalks prey, the bell rings before she can pounce, giving the bird or mouse an audible warning.

3 But cats often then show their natural cunning: they adjust their hunting strategy. They start to move even more slowly than usual to stop the bell from ringing as they close in on the target.

Recent estimates suggest that 55 million birds are killed a year in the UK by domestic cats. However, it appears that they mainly take sick or old birds that were likely to die anyway, and so their effect on the overall bird population is not necessarily a cause for concern.

**Collar safety**

Be sure to choose a safety collar to keep your pet safe. This will break apart if your cat becomes trapped by her collar, as can happen when she is climbing. If this didn't happen, she would be at risk of being strangled by the collar.

**Below:** Although cats usually stalk prey on the ground, they can jump up and knock down a small airborne bird with their front paws.

**Right:** Cats like to explore, and they are able to climb trees easily. They may even hunt up there, off the ground.

★ Your cat is happy to wear a collar with a bell attached

★ ★ When she is hunting, the bell rings when she moves

★ ★ ★ She has learned to move without the bell ringing

# Rules of the Road

Plan ahead to try to keep your cat safe from the threat posed by road vehicles. Take preventive action (as described opposite) if your pet is allowed to go outdoors. Unfortunately, their wandering lifestyle does make cats susceptible to this risk, especially if they're allowed out to roam at night.

1 Cats have no road sense and may run out in front of approaching vehicles without warning. This situation is even more dangerous at night, as they are then harder for motorists to spot.

2 If your cat normally wears a collar, fit a reflective strip to it. This helps to alert a driver to your pet's presence in the road before it is too late.

Sadly, many cats are injured or killed on the roads every year. Neutering can help to keep your cat safe from this hazard as it reduces the cat's desire to roam away from home.

**3** Should your cat be involved in a collision, arrange a rapid veterinary checkup. Cats can suffer serious and potentially fatal internal injuries that may not immediately be obvious.

★ Your cat sometimes wanders near or across the road

★ ★ Your pet comes to you when she is called

★ ★ ★ She comes indoors readily in the evenings

**Above:** Cats can be attracted to cars by the warmth of the engine, jumping up and sitting on the hood.

**Below:** The relatively warm road surface can attract a cat, leading it to curl up and go to sleep on the tarmac.

### Cats' eyes
Bright headlights can easily dazzle a cat. This is because of a mirror-like, reflective lining called the *tapetum lucidum* which is present at the back of each eye. It normally helps cats to see well in the dark.

# Cat on a Leash

You might get a few odd looks, but more and more cat owners are taking their pets for walks on a leash. Not all cats enjoy the experience though, so it helps to teach your cat to accept a harness and leash from an early age. The leash should attach to the back of the harness, not around the neck.

1 Kittens are easier to train to walk on a leash than older cats. Once your cat is comfortable wearing the harness around the home, attach the leash.

2 Never let your cat out just wearing her harness as it could become caught up on branches or fencing, and she may not be able to free herself.

Keep the leash fairly tight when walking down a street, so that your cat can't dash out into the road.

**3** The Siamese and Oriental breeds have a reputation for being the easiest to train to walk on a leash. Use tasty treats to reward good behavior.

### Steer clear of dogs

If you are out in the street or park, watch out for dogs. Any close encounters with them are likely to upset your cat and, when frightened, she may become almost impossible to handle or to calm down.

Many indoor-dwelling house cats get insufficient exercise to ward off obesity. If you are worried about letting your cat roam outside, walking with her on a leash is a great way of allowing your pet to get some welcome exercise.

★ Your cat struggles when the harness is first put on

★ ★ She is happy to walk around the house in the harness

★ ★ ★ Your pet walks happily and well on the leash

# Lost Kitty

Cats can disappear for a variety of reasons. Whatever the cause, if your pet goes missing this will be a very worrying period. Try to minimize the risk by planning ahead. First and foremost, make sure that your cat has been microchipped so that she can be easily identified if she gets lost.

1  Neutering is the best way to stop your cat straying, and it has other benefits too. It makes tom cats much more pleasant living companions, and ensures a female doesn't produce a litter of unwanted kittens.

2  Keep an eye on the weather forecast and calendar. Loud noises and flashes of light, as you get with thunderstorms and fireworks, can be very frightening for cats. Train your pet to come indoors when she is called.

Always keep your cat indoors for at least two weeks after a house move, to stop her straying.

**True or False?**
A cat can run over short distances at a speed of roughly 30 miles/hour.

☐ True
☐ False

**Above:** Microchipping can help to reunite you with your pet if she strays. The microchip—about the size of a grain of rice—is inserted under the loose skin in the neck region as shown here.

**Right:** Detecting the microchip is easy with a special reader. This activates the chip, detecting its unique ID number. Don't forget to update your details on the database if you move!

### Search party
If your cat does suddenly go missing, don't automatically fear the worst. Many cats simply get locked in somewhere. Ask your neighbors to search in their outbuildings, as cats can wander into sheds or garages and be unable to get out again when the door is closed.

# Going Walkabout

Cats can wander far and wide in their neighborhood, and yet they manage to find their way home without getting lost. Even so, bear in mind that if you move into a new home that is just a short distance from where you were previously living, your cat may decide to pack up and disappear back there!

1 Once your cat has settled into a new home with you, let her start venturing outside for relatively short periods.

2 Don't force your cat to go outdoors at first. The surroundings will be strange, and cats are actually quite nervous by nature.

3 Be prepared to go out with your pet into the yard, and play there, so your cat can build up her confidence.

Young cats such as this Siamese can be allowed out to explore once they have completed their course of vaccinations.

### High density dwelling

Problems often arise because domestic cats live in much greater population densities in towns alongside us than they do in rural areas. The close proximity of other cats may lead to fierce battles over territory.

★ Your cat follows you out into the yard

★ ★ She explores the yard on her own

★ ★ ★ Your pet comes in obediently when she is called

**Above:** Another advantage of neutering is that it reduces the likelihood of your cat becoming involved in a serious territorial fight with a rival cat.

**Below:** Although generally sure-footed, cats—and particularly kittens—can fall from tall buildings. Ensure your balcony is safely enclosed.

# Home From Home

When roaming outdoors, before long your cat is bound to venture farther afield into neighboring yards. This may lead to her getting to know your neighbors, although you may be unaware—at least initially—of this developing friendship. So it pays to keep an eye on where your cat is going.

1 Things can easily change. The neighbor may start offering your pet some tasty household scraps, which taste better than her regular food, so your pet starts to visit regularly.

2 Your cat carries on building a bond with your neighbor, perhaps venturing indoors and spending time there. This is most likely to happen in a house without a resident cat.

3 Before you know it, your cat is disappearing for long periods. This is likely to be worrisome, and she may seem to be off her food when at home, because she's getting fed elsewhere.

Cats like attention, so always make a fuss of your pet to deter her from straying.

## Eating out

Keep a check on your cat's weight. This is important for your pet's general health, with obesity among cats being on the increase. Weight gain can also alert you to the fact that she is being fed somewhere else.

★ You suspect your cat may be getting food elsewhere

★ ★ She wanders off from home but always returns

★ ★ ★ Your pet generally eats well at home

Prepared cat foods today are far more palatable than they used to be. However, if your cat starts to seem picky about food, you may need to try different flavors or even change brands.

# The Final Score

So how is your cat doing when it comes to assessing her role in the world, and how she relates to her environment? Having a well-adjusted cat helps to ensure that you too can have a more relaxed lifestyle.

Record how your pet is progressing on the special star chart opposite. Put the number of stars scored in each quiz into the boxes. For "True or False" questions score three stars for the correct answer but none for the wrong one. Add up the total number of stars your cat has scored in this quiz—you will need this for the final score chart on the opposite page. When all the results are in, you will be able to work out just how smart your pet really is!

Cats can prove to be very resourceful and adaptable creatures. Good owners can harness and extend these natural skills to fashion a pet that is both smart and fun to be with.

## Farther Afield

Pages 172–173
### Going Somewhere?
☐ Number of stars

Pages 174–175
### Walk on the Wild Side
☐ Number of stars

Pages 176–177
### Rules of the Road
☐ Number of stars

Pages 178–179
### Cat on a Leash
☐ Number of stars

Pages 180–181
### Lost Kitty
☐ True

Pages 182–183
### Going Walkabout
☐ Number of stars

Pages 184–185
### Home From Home
☐ Number of stars

## Adding Up the Stars

Now it's time to add up the scores for all the stars your cat has achieved in the charts throughout the book. In each case, did they get mostly 1 star, 2 stars or 3?

Page 76
### Chart 1    ☐ ☐ ☐  Number of stars

Page 103
### Chart 2    ☐ ☐ ☐  Number of stars

Page 141
### Chart 3    ☐ ☐ ☐  Number of stars

Page 169
### Chart 4    ☐ ☐ ☐  Number of stars

Page 187
### Chart 5    ☐ ☐ ☐  Number of stars

### Total    ☐ ☐ ☐  Number of stars

**How did my cat score?**

Mainly 1 star = your cat is smart, but may need some work

Mainly 2 stars = your cat is becoming a brainiac

Mainly 3 stars = your cat is a gold star pupil

**Is my cat improving?**

| | |
|---|---|
| First total star score: | Date: |
| Second total star score: | Date: |
| Third total star score: | Date: |
| Fourth total star score: | Date: |
| Fifth total star score: | Date: |

# Index

# Picture credits

Quercus

New York • London

Copyright © 2015 Quercus Editions Ltd
First published in the United States by
Quercus in 2015

Any member of educational institutions
wishing to photocopy part or all of the work
for classroom use or anthology should send
inquiries to permissions@quercus.com.

ISBN 978-1-62365-489-4

Library of Congress Control Number:
2014948254

Distributed in the United States and Canada by
Hachette Book Group
1290 Avenue of the Americas
New York, NY 10104

Manufactured in China

10 9 8 7 6 5 4 3 2 1

www.quercus.com

Text by David Alderton
Edited by Philip de Ste. Croix
Designed by Sue Pressley and Paul Turner,
Stonecastle Graphics Ltd